The Moon Ga

A biodynamic guide to getting the best from your garden

Peter Berg

TEMPLE LODGE

The Moon Gardener

A biodynamic guide to getting
the best from your garden

Peter Berg

Contents

Introduction: Living earth

Dear readers

I have met all kinds of people during my 40-year professional gardening work: a large number of keen amateur gardeners who have asked me countless interesting questions; the international trainees I have introduced to biodynamic practice; and not least participants at the annual gardening course 'Practical Gardening with Peter Berg' which is held at the Demeter training and show garden in Binzen, Germany.

Partly through all these encounters I gradually formed the idea of writing a book on the foundations of biodynamic practice. I was concerned to explain to lay people in clear, accessible language the methods that professional gardeners use as a matter of course, and help them to do the same.

In the great cosmic context in which we are all embedded, the moon is an important – if relatively small – mover and impulse-giver.

My great thanks go to Maria Thun, the grande dame of biodynamic gardening, and her son Matthias, for 35 years and more of intensive dialogue about practical ways of using and working with the constellations in agriculture and horticulture, and deepening this practice. They have elaborated the foundations of this work in many scientific trials.

I am very pleased and grateful that I had the good fortune to meet Maria Thun in person as a young man, and to incorporate much of her experience into my daily work as head gardener. Her special credo has always been, 'Don't make my approach (with specific days for different tasks) into a dogma.' I have always kept that in mind, and it is worth remembering for your own domestic gardening — it makes no sense to try to use the 'right' constellation in adverse weather conditions.

I hope, dear reader, that your enthusiasm for gardening will gain much impetus and inspiration from this book. The cosmos has much to offer us if we give it the chance to unfold its nurturing work through an enlivened soil and the sensitive world of plants. By this means, indeed, food can become a kind of medicine. The earth needs support through our use of the preparations which Rudolf Steiner described in 1924, in order to counter environmental pollution and continue to maintain its fertility for future generations. May this endeavour succeed through our common efforts.

Peter Berg

The Foundations

Inspiration from Rudolf Steiner

Immediately after the First World War, the first chemical fertilizer was introduced into agriculture in the form of saltpetre – in other words gunpowder. It was used to add nitrogen to the soil, and played an important part in the changes that took place in agriculture from the mid-nineteenth century onwards. Increasing industrialization, the use of new machines, and chemical substances in the form of fertilizers, herbicides and pesticides, allowed agricultural yields to grow. Farmers' work grew simpler. Yet only a few years after the introduction of chemical fertilizer, attentive farmers started to notice worrying changes in their crops: seed quality declined, and the plants themselves became more susceptible to environmental factors, diseases and pests.

The birth of organic agriculture

During this period, Rudolf Steiner (1861–1925) was giving lectures throughout Europe and, as philosopher and founder of anthroposophy, publishing his ideas on the most diverse themes. His approaches to such fields as education, special needs education, medicine and all forms of art offered innovative approaches. His audiences included farmers, and some of them asked Rudolf Steiner to support their efforts at countering the declining quality of seed. At Whitsun 1924 this support was forthcoming in the form of a conference entitled 'The Spiritual-scientific Foundations for a Flourishing Agriculture'.* In the ten days of this event, and the eight lectures given there, Rudolf Steiner laid a comprehensive foundation for a renewal of agriculture. The conference became the cornerstone of biodynamic agriculture, and at the same time marked the birth of organic farming methods throughout the world.

Around one hundred people attended the conference at the Koberwitz estate of Count Carl von Keyserlingk in Upper Silesia, near Breslau, with its 4000 or so hectares of agricultural land. Among them were various managers of large estates in the east of the country, for Steiner's ideas were also intended for large agricultural concerns of several hundred to several thousand hectares – estate sizes that were by no means unusual in those days. One can see how great was the participants' interest in the theme from the fact that it was not in the least easy for farmers to leave their fields at

---

* See *Agriculture Course* by Rudolf Steiner (Rudolf Steiner Press).

this season of the year, at the busy time of hay harvest.

To be able to follow Rudolf Steiner's explanations, participants were required to have read, as advance preparation, fundamental writings by him such as *Occult Science* and *Knowledge of the Higher Worlds*, since his lectures drew on the great interconnections between cosmos and earth. He also gave many specific, practical recommendations for diverse aspects of agriculture, including ideas for beneficially shaping the landscape through such things as hedge-laying and forestry.

Ecosystem integrity: the individuality of the farm

At this conference Steiner also first introduced his concepts of 'farm organism' and 'farm individuality' – two key ideas in biodynamic agriculture. They define an agricultural operation as, ideally, a complete, autonomous organism which can itself produce everything needed for growing high-quality, healthy food. Thus the farm is not reliant on buying in things from outside, such as seed, fertilizer, etc., which may impair quality. Instead, it is a highly developed, sustainable ecosystem. To maintain the soil's fertility, therefore, the 'farm individuality' requires animal husbandry – above all cattle – alongside vegetable, fruit and grain cultivation.

Homoeopathy for the soil

From today's perspective, Steiner's recommendations for the so-called 'preparations' must be seen as especially outstanding in value. Put very simply, these are 'homoeopathic medicines' for the soil, compost and plants. Among other things they lead to increased micro-organism activity in the soil, enlivening it and thus helping produce healthy, strong plants.

In principle, the preparations are similar to the substances which Steiner proposed for human medicine, and which at that time, and still today, are produced by the companies Weleda and Wala.

Rudolf Steiner's directions for making the agricultural preparations are far-reaching. To fully understand the processes involved one requires not just a knowledge of chemistry and geology but also wide-ranging insight into Steiner's ideas. By directing people's attention to the cosmos, in particular the zodiac and the planets in the earth's immediate vicinity, Steiner was urging them to engage with a breadth of cosmic influences in the most down-to-earth ways.

Demeter

As members of the experimental group founded in 1924, farmers initiated a lively culture of experimentation and dialogue. In 1927 the first cooperative was founded to market biodynamic produce, and a few years later this was renamed as the Demeter Association. The name 'Demeter' is still used today as a trademark. It is the name of the ancient Greek goddess of fertility and motherhood, whose activity manifested in the fertility of the earth, corn, seed and the seasons, and who according to legend taught human beings how to cultivate crops and handle the plough.

Practical dialogue

Rudolf Steiner repeatedly said that all his suggestions must be thoroughly tested in agricultural practice. His audience was therefore urged to try them out in their farms and gardens, and engage in dialogue with others about the results. To facilitate this, the farmers founded a biodynamics experimental group during the Agriculture conference in 1924. In those days this was a major task, since participants' farms and estates were spread all over Germany. Dialogue, discussion and meetings required considerable time without the modern means of communication available today. These pioneers took on the task, nevertheless, with great energy and commitment.

Biodynamics prohibited

Despite more difficult conditions under the Nazis, up to about 1941 the method was able to develop in Europe, with increasing numbers of farmers using it. Especially in Germany, Switzerland and Austria, large areas of land were cultivated according to biodynamic principles.

From 1941, the Nazis put such adverse pressure on the whole anthroposophical movement that work could only continue in secret, and at great risk to life and limb. A good number of farmers relinquished the approach, and many never returned from the war.

The cosmos works into plants via the zodiac signs, moon and planets. Here, at the market garden's show garden, the zodiac is represented through appropriate plants and natural materials.

After 1945, therefore, the initiative lacked leadership.

Important estates that had been farmed biodynamically during the pioneer period were located in eastern Germany. In socialist countries and the GDR, these were turned into collectives where the biodynamic approach was prohibited. Some survivors of this period however were able to take up their work again in West Germany.

Early pioneers

Two important individuals from the biodynamic research group, Ehrenfried Pfeiffer and Maria Thun, should be mentioned here, since their work has been a source of inspiration for me. Ehrenfried Pfeiffer (1899–1961), a pioneer of biodynamic agriculture and a pupil of Rudolf Steiner, contributed decisively to establishing the method, and developed it through his international advisory work. In the USA in particular he inspired much activity.

Maria Thun: research into the workings of moon and planets

After the Second World War, Maria Thun emerged as a leading figure in the biodynamic research group. She took up Rudolf Steiner's suggestions and, over decades, systematically undertook countless trials on plants – partly in collaboration with Giessen University.

She found that the same plant (e.g. radishes) will develop in different ways depending on which day the ground is prepared for sowing or cultivated by hoeing, and in addition that storage and use of crops after harvest is likewise differently affected.

These findings led to her developing the concepts of 'root, leaf, blossom

Strawberries, classified as 'fruit plants' by Maria Thun

and fruit days'. It is thanks to her research that we use these terms today as a matter of course in biodynamic work — although arduous effort was needed to get there.

Radishes
belong to the root plants

Tagetes (marigold) form
attractive blossoms and are
therefore assigned to
blossom plants.

All types of lettuce belong
to the group of leaf plants

Zodiac signs support plant types

Maria Thun's research showed that cosmic rhythms underlie plants' different ways of growing. For instance, radishes always developed most typically, i.e. best, when the moon was in front of the zodiac signs Bull, Virgin or Goat at sowing. Based on these and other findings, she coined the term 'trigon' to describe the fact that three respective zodiac signs have a positive effect on each type of plant:

- For leaf plants (e.g. lettuce), the star signs Fishes, Crab and Scorpion (water signs)
- For blossom plants, the trigon Waterman, Twins and Scales (air signs)
- For fruit plants the star signs Ram, Lion and Archer (fire signs)
- For root plants, Bull, Virgin and Goat (earth signs)

Sidereal moon cycle

Maria Thun's research is based on astronomy and is oriented to the sidereal moon cycle lasting 27.3 days. Seen from the earth, after this period the moon has returned to its original

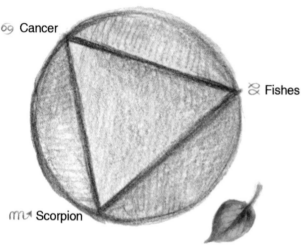

♋ Cancer

♓ Fishes

♏ Scorpion

Leaf trigon (water)

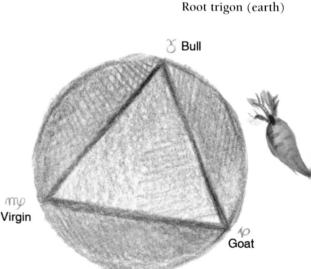

Root trigon (earth)

♉ Bull

♍ Virgin

♑ Goat

position in relation to the fixed stars. In its orbit around the earth, it crosses the zodiac signs within a sidereal month, passing in front of a new sign every two to four days, depending on the relative size of the sign, and mediating its influences to the earth via the elements of earth, water, air/light and fire. This is why, in her research, Maria Thun does not use the classical segment-division of 30° but works with ratios corresponding to the star signs' actual different sizes in the heavens, whereas astrology divides the firmament into twelve equal parts.

Weed tip from Maria Thun

Maria Thun has also investigated 'weed behaviour', and her findings are a great help in deciding what methods to use and when. For instance, it is possible to get weeds to germinate more strongly at a particular constellation (moon in front of Lion), and then suppress development of the germinated weeds when the soil is next cultivated – around two to three weeks later – at a different constellation (moon in front of Goat).

Patience and time

The following is the most important principle. The best effect from the cos-mos can only be expected with good seed quality (biodynamic or organic seed) and a living soil, in harmony with

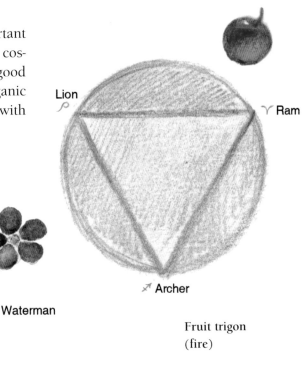

♊ Twins

♎ ales

♒ Waterman

Blossom trigon (air)

Lion
♌

♈ Ram

♐ Archer

Fruit trigon (fire)

In around 27 days the moon orbits once around
the earth, passing across the twelve zodiac signs
within the sidereal moon cycle. The star signs
are divided into segments of differing sizes in
astronomy and in Maria Thun's approach,
whereas astrology divides them into equal-sized
segments.

18

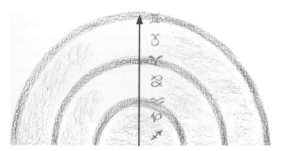 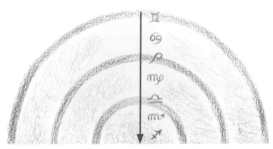

Ascending moon: the moon rises from the star sign Archer to the star sign Twins. This is a poor time to sow and plant, but gardening work such as pruning and harvesting can be carried out.

Descending moon: the best time for planting is when the moon is descending from the star sign Twins to that of Archer. This is what we call 'planting time', when new planting, replanting and planting out can best be done, as well as sowing.

the right application of biodynamic preparations, and cultivation work carried out at the right constellations for a particular plant type. Don't be downcast if you fail to notice significant differences immediately. It takes time — both for the soil and for oneself too. In my own case, it took years for this approach and these insights really to become second nature to me.

Ascending and descending moon

As already mentioned, the moon orbits once around the earth in around 27 days, thus passing across the star signs. At the same time it ascends and descends in relation to the ecliptic, moving from its deepest point in the star sign Archer to its highest point in Twins. This means that each night the moon describes a somewhat higher path (ascending moon) or lower one (descending moon) in relation to the fixed stars than the previous night. (In the southern hemisphere, this is reversed.)

When the moon is in its descending phase, the sap is falling in plants and all forces pass to the root. The earth 'breathes in'. This is therefore the planting period.

When the moon is ascending, the plants' saps rise once more into the upper parts of the plant. The earth 'breathes out' and forces of growth above the soil are intensified. This phase is called non-planting time.

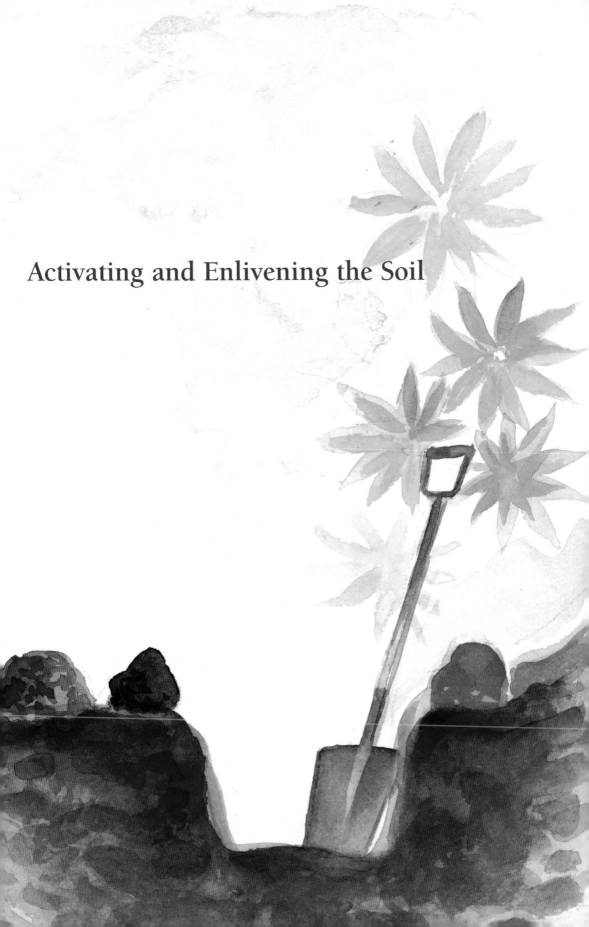

Activating and Enlivening the Soil

The garden's position

For plants to grow as well as possible, we first need to make a careful analysis of the garden plot itself. Its position and other external factors will affect how cultivation thrives, and it is therefore important to be aware of them.

Being open to cosmic impulses which influence plants via the soil.

Sloping or level plot: erosion

Is my garden level or on a slope? This will determine how the garden is arranged. On a sloping plot it is important for beds to be laid as far as possible crossways to the slope, due to the risk of erosion.

Microclimate: frosts

If your garden lies in a depression, you should reckon with early and late frosts since this will draw in cold air. Or is the plot in an open, level setting? This will mean that moisture dries more quickly because the wind 'whistles' over it. If the garden is on raised ground, it is likely that, as in a declivity, it will be exposed to late frosts in spring and early frosts in autumn.

Position: sun and shade

Is your plot, or parts of it, exposed to sunlight throughout the day, or are there times of day when it lies in shade? One needs to consider what plants cope with such positions. Shade can also, of course, be cast by buildings outside your own garden. If that is the case, they may also protect the garden from strong winds, which will be pleasant for both you and the plants. Large trees in or adjoining the garden cast strong shade.

Overwintering crops such as spinach and lamb's lettuce are hindered by leaf-fall.

When dividing up the garden, one should also take account of the long branching underground roots of trees, which will 'compete' for nutrients with your cultivated plants. Leaf-fall in autumn is also worth considering. Spinach, lamb's lettuce and all over-wintering plants planted in autumn will be severely hindered by leaf-fall.

Soil: groundwater and moisture

Groundwater level in the garden affects the soil's basic moisture content. The higher the groundwater, the damper the ground. One way to discover the relative groundwater level is to talk to your garden neighbour, for instance about his experience of digging a well. You may only be able to reckon on a little water rising from below through the soil capillaries (hair-fine, inter-connected, vertical soil interstices through which water rises by capillary action). In that case hoeing should be undertaken to loosen the upper surface of the soil. By loosening the soil in this way, the upper ends of the soil capil-laries can be broken so that the water rises without being blown away at the surface by the wind. Thus less water evaporates and the gardener needs to water less.

It is always worth talking to your neighbour or the gardener's previous occupant, who will often have had years or even decades of experience with the conditions there, such as weather and the soil's response to it.

Knowing your soil

Besides the external factors mentioned, it is very important to know your garden soil very well, both its type and quality. One has to develop understanding of the soil's nutrient network as a whole, for plants and soil are in living interaction, continual giving and taking. Every gardener should be aware of the type of soil he and his plants are working with. Only then can he make the best of what is available — since he can't, after all, pick up his garden and put it down elsewhere.

Soil profile

Soil is the result of long processes of geological and climatic development, as well as of human influence on the various underlying rock structures such as granite or limestone. By making a profile of the soil we can discover the prevailing type of earth or soil in our garden.

To do so, dig a hole roughly one metre deep (in dry conditions a little deeper) to see the soil strata clearly. While digging you can also tell how much resistance you meet, which gives an idea of the efforts plant roots will need to delve into it.

Soil types

There are four types of soil, described below.

Sand and gravel

Sand and gravel soils are often found in floodplains and old river beds, for instance in the Rhine basin. They are often light, loose soils that are easy to work. Such a soil means you can start gardening work early in the year, since it dries off quickly. However, it will not have good water retention, and so the gardener must water or hoe it more often. This is true particularly in the warm, dry summer months when the ground dries out more quickly.

Another characteristic of sand and gravel substrates is that they warm up more quickly than other soil types. In spring, therefore, one can often start early with the first gardening jobs. In summer, however, lower capillary action means you are likely to have a lack of moisture. Mulching well and regularly will help to improve soil water content.

Limestone soils

A calcium-rich soil is light in colour and sticks to your hands. In dry conditions it tends to form clumps, and does not warm up as quickly as other soils. It is usually hard to work, for in spring the earth on the surface will already be dry while below it is moist and sticky. A soil of this type can be improved by careful use of compost.

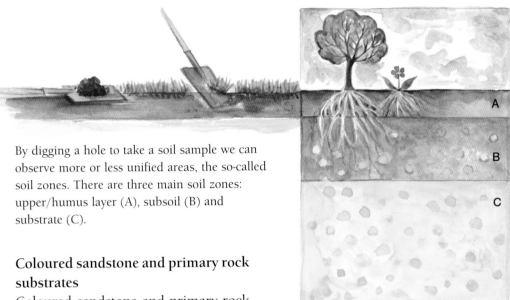

By digging a hole to take a soil sample we can observe more or less unified areas, the so-called soil zones. There are three main soil zones: upper/humus layer (A), subsoil (B) and substrate (C).

Coloured sandstone and primary rock substrates

Coloured sandstone and primary rock substrates usually give the soil a reddish colour, and the fine sand content makes them easy to work. The amount of clay affects water retention capacity. A high clay content means water is well retained. At the same time, though, the soil is less permeated by air and it becomes heavier due to higher density – requiring greater effort by the gardener.

A soil of this kind is easiest to work in a relatively small 'window' of a few hours each day.

Loam sediments

Loam is one of the most fertile soils in the world, and is found on all continents. It is formed of airborne dust sediments carried by the wind. The loam deposits in Central Europe date from the Pleistocene era.

Besides good water retention, loam has other positive qualities. As it is composed of the most varied minerals, the substrate is also extremely rich in nutrients. The low clay content means it is a loose, easily worked soil.

Loam is the dream of every gardener, for it brings good yields. It is found primarily on sloping plots and in floodplains.

A deeper perception and analysis of the soil

There are various ways of discovering the condition of your soil. The better your knowledge of it, the easier you can choose from a wide range of soil-improving measures.

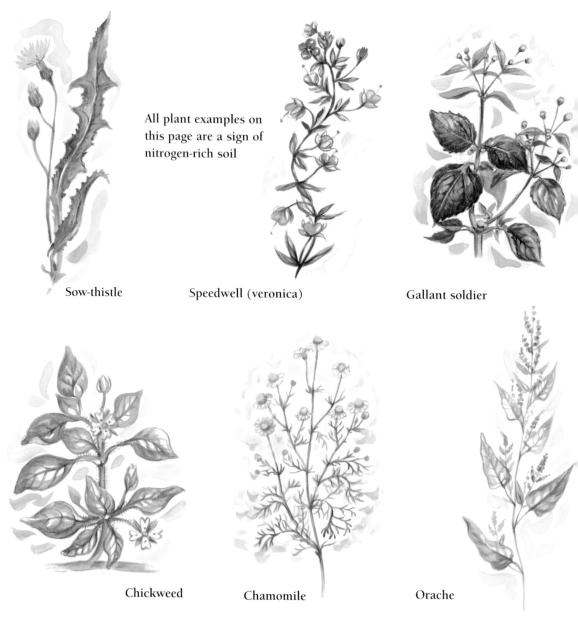

All plant examples on this page are a sign of nitrogen-rich soil

Sow-thistle

Speedwell (veronica)

Gallant soldier

Chickweed

Chamomile

Orache

Indicator plants

Plants or weeds growing wild in a garden can give an important indication of soil condition and are therefore also called indicator plants. The following table shows some common indicator plants with their significance for soil condition.

Dandelion

Soil indicator plants

Soil condition	Indicator plants
Nitrogen-rich soil	Stinging nettle, red dead-nettle, orache, chickweed, cleavers, common sow-thistle, petty spurge, shepherd's purse, speedwell, fat hen, gallant soldier, ground elder, cocksfoot grass
Nitrogen-poor soil	Wild carrot, stonecrop, clovers, wild radish, hairy vetch (hairy tare)
Acidic soil	Sheep's sorrel, feverfew, wild mint, haresfoot clover, rosebay willowherb
Alkaline soil	Common toadflax, field pansy, charlock, creeping cinquefoil, meadow cranesbill, hoary plantain, ragwort, silverweed
Damp or waterlogged soil	Dock, lady's smock, meadowsweet, wild mint, creeping thistle, creeping buttercup
Dry soil	Common storksbill, hoary plantain, dovesfoot cranesbill, yarrow, ragwort
Compacted soil	Silverweed, ratstail plantain, rayless chamomile (pineapple weed)
Shady site	Ground elder, ground ivy, enchanter's nightshade, nipplewort

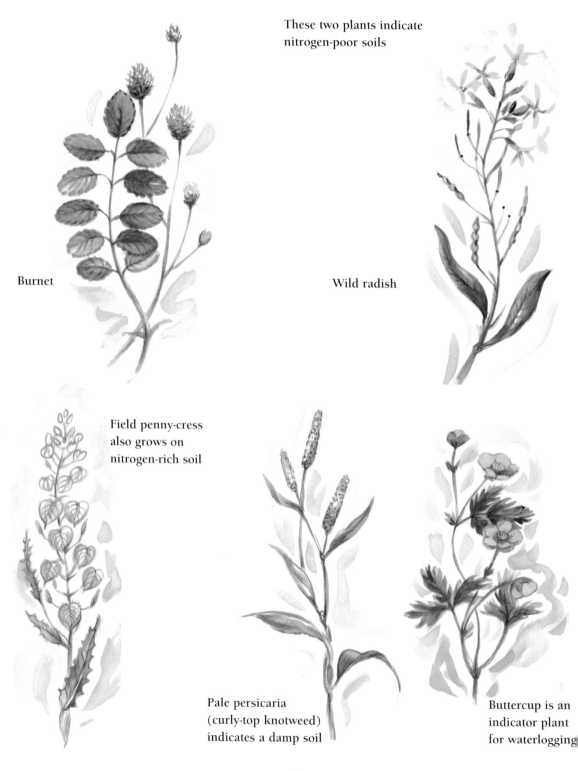

These two plants indicate
nitrogen-poor soils

Burnet

Wild radish

Field penny-cress
also grows on
nitrogen-rich soil

Pale persicaria
(curly-top knotweed)
indicates a damp soil

Buttercup is an
indicator plant
for waterlogging

Feverfew indicates acidic soil

Common poppy indicates sandy soil

The finger test

The finger test can tell you about your garden soil. Rub the earth between your fingers and ask: How does the soil feel? Is it sticky, moist or even wet? Is it finely structured? How much sand and how many stones does it contain? Does it have unrotted organic constituents, e.g. bits of cabbage from last year? (This would be a sign that the soil is not able to 'digest' organic substances quickly.)

Take a small sample of the soil and try to roll out a pencil-thick sausage with it. Adding a little water will make the sample more malleable. Can you manage this or not?

- If you can't, the soil belongs to the sandy soils
- If you can, it belongs to the sandy loams and clays

The crumble test

This will tell you what proportion of fine earth the soil type contains, and thus whether your soil is more sandy or loamy. The crumble test is not as precise as the finger test but will give you a feeling for your soil. From a spadeful of earth with the vertical soil profile take roughly a handful of soil, noting whether you take it from higher up or lower down the profile. Remove all plant parts and also roots. The earth mass should be neither too wet nor dry. Take a little earth in your hand and press this briefly together, then open your hand again and note what happens:

- The soil trickles through your fingers – the soil contains up to 5 per cent sand
- The soil crumbles and disintegrates through your fingers – the soil contains 5–40 per cent loam
- The soil can be shaped and single slits arise when pressed together – the loam content is 40–50 per cent
- The soil can be formed into sausage shapes – it has up to 50 per cent clay content

The handful of humus allows you to tell many things about your soil

The smell test

The smell of the soil also gives important indications of its condition. Does it smell good, like forest earth? Then it is good quality, for forest soils have the best microbial milieu.

If the garden soil is black and smells boggy, this is a boggy soil. Typically it will contain plant parts as fibrous substance.

If it smells mouldy, sulphurous or of rotten eggs, this is usually a sign of poor decomposition of organic substance. It is usually caused by lack of aeration in a compacted soil, e.g. animal manure dug in the previous year. In this case you will need to loosen the soil by one of the many means described on pages 36–37, to improve the tilth.

The spade diagnosis

This method is easy to use and gives a snapshot of the soil's condition, showing what scope there is for plant roots to establish themselves.

Dig a spade into the earth and lift out a roughly tile-sized piece, then place the spade with the soil on a table or directly

The cross-section from the spade sample shows the soil profile clearly.

In the spade diagnosis, the soil is lightly scraped apart

on the ground. Now observe the soil sample and lightly scrape it apart. Note how it disintegrates and what crawls out. If the soil seems crumbly throughout, and full of creatures, it will be good for plant roots. In such a soil, for instance, carrots will grow well (rather than becoming leggy). The more crumbly the soil, the easier a time plants roots will have. If the soil seems solid and impermeable, on the other hand, careful application of compost will be needed.

Counting earthworms

Counting earthworms will give further insights into the vitality of the soil and your garden's potential fertility.

The earthworm is one of our most important visible helpers. It forms earthworm deposits on the surface of the soil, but also in sub-surface cavities. The wormcasts arise when the worm mixes earth and organic constituents together in its digestive tract, adding its digestive mucus and excreting everything again. The excreted substance is the basis for soil organisms that are important helpers on the path towards greater soil fertility. They further digest the earthworms' excretions. The mucus secreted by worms in their holes so that they can glide along with their tiny bristles is

The soil is alive: from woodlice to slugs and snails

If you dig your spade into the ground and remove a small sample to the spade's depth, you can already see soil creatures with the naked eye: springtails, mites, small bristle worms, beetles and beetle larvae, centipedes, ants, woodlice, fly larvae, spiders, earthworms, slugs and snails. This list is in ascending order of size. A large number of the most diverse creatures shows a soil's vitality.

To count earthworms, work over an area in the garden of about one square metre, digging down about 30 cm (the depth of a spade). Do not carry out the earthworm count during a dry period since earthworms will then withdraw deeper into the ground (up to 80 cm).

The earth you dig out is placed on a tarpaulin, where you can count the earthworms and their eggs at your leisure. If there are around 80 worms, including eggs, you have a moderately vital soil whose activity will need to be intensified through appropriate measures and treatments.

Around 150 worms with eggs shows your soil has good organic activity. Finding 250 or more indicates a level of vitality that can scarcely be improved. With such a soil, the work you do will primarily be geared to retaining this high level of vitality.

Already at spade depth one can see a soil's vitality in the form of many small creatures.

also an important substance for soil organisms.

Chemical analysis

To take a reliable sample, you will need to remove soil from at least ten to twelve places at differing depths (between 0 and 30 cm). This sample is now mixed together well, and around 500 g is taken straight to the laboratory (see page 123). There you can ask for various determining parameters, for instance:

- Calcium content from pH measurement. The optimum pH value (soil acidity) is between 6.5 and 6.8.
- Total nitrogen content including nitric nitrogen. Only nitric nitrogen is immediately available to feed plants.
- Phosphate and potash as further chief nutrients

- Trace nutrients, especially magnesium
- Organic substance, so-called humus content

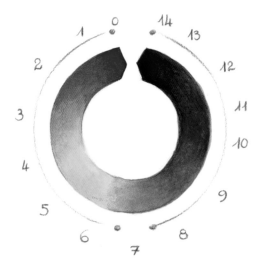

The pH value is the degree of soil acidity. It tells us whether the soil is acidic, neutral or alkaline. Most cultivated soils have a pH value between 4 (strongly acidic) and 8 (strongly alkaline). The acidity of the soil decisively affects absorption and availability of nutrients for plants.

Soil tests at a glance

Finger test: reveals the nature of the soil – sand, loam or clay.

Crumble test: likewise shows the nature of the soil (sand, loam, clay, mixture) but not as precisely as the finger test.

Smell test: shows the condition of the soil – boggy, mouldy or fresh – and whether soil treatments such as loosening of the ground or tilth improvement need to be undertaken.

Spade diagnosis: snapshot of soil condition. Does the soil seem permeable, crumbly and alive, or heavy and impermeable? In the latter case, applications of compost can help.

Counting earthworms: indicates the soil's fertility and vitality. Soil improvement required below 80 worms/eggs. Over 250 worms/eggs indicate a very fertile garden soil.

Chemical analysis: assessing pH value, chief and trace nutrients, and humus content.

The soil speaks

I received the following text from an old friend in the biodynamic movement, Eugen Burnus, who is now about 80. He has spent his life working in the interests of the soil in various regions of the world.

I hope these lines will help us to reflect on the way we relate to the soil we walk upon each day or drive over in our vehicles.

A pause for thought

Christmas 2009

They are celebrating your birthday once again. Do they know that you came to earth in pure love, in the same way that the whole of Creation arose in pure love? You do not wish to redeem human beings from supposed sins. Your being is embodied only in love for all creatures.

I, the good earth, know about this mysterious power of love. I was born through the love which you give to all people as a gift. I am not nature, I am the culture of human love. I changed from forest to fertile fields in a process lasting centuries. Generations lived for me and with me, and perceived me. In this shared life, full of mutual dedication and love, I was able to thrive and give human beings the gift of rich harvests. But this wonderful time is past. The number of people who lived with and for me has continually dwindled. In the place of human beings and their love came machines. People had ever less time to speak with me, to perceive me. They were less and less able to work upon me meditatively, in reverence. They showed ever less regard for my life.

And so, from being man's partner I became his slave, without love or perception of my own, rich being. Rich harvests were ever less possible to give. But they found ways to exploit me with subtler methods. By all means I was compelled to provide the fruits of my life, and a culture of centuries faded.

But to my surprise, new people came from the cities and sought to live in community with me. They had as yet no experience of my needs and the necessary rhythms. Nor did they as yet have the ears to hear me. I kept trying to tell them. Sometimes I succeeded in reaching their dreams. And so today I live full of hope when I see how many are coming to me and learning to live rhythmically with me again, to perceive me in love and meditation. Only when this happens day-in, day-out can I once again live as the partner of human beings, and supply them with healthy fruits. Only the shared responsibility of all human beings for me as sustaining soil will allow me to live fruitfully.

Through centuries I was built up by love, and have almost starved to death through lovelessness. The love of human-

*kind can transform me
again, and in love I will
be able to give them rich
harvests. Please do not leave
me alone! Dance and sing
with me, and transform the
work with and upon me into
a healing and holy meditation.*

Giving good things to the soil

Lupins belong to the family of Leguminosae and can add atmospheric nitrogen to the soil, thus making a major contribution to soil fertility.

An old gardener's saying holds that 'a good harvest is the best preparation for the next crop'. The insights gained from diagnostic methods we have so far examined can provide a range of ideas for soil improvement which we will consider in the further course of the book. They are briefly outlined below.

Adding compost

A high-quality compost leads directly to improvement of soil quality. How do I make this? (see 'Compost', pages 51–65.)

Crop rotation

A carefully planned, well-designed and implemented crop rotation serves to continually improve and protect the soil. This is not a short-term method for short-term results. The soil, as living organism, must be allowed time to respond, and only then will the garden's ecosystem remain in balance (see the chapter on Crop Rotation and Green Manures, pages 67–79.).

Green manures

Green manuring is another method of active soil care. By planting special green manure plants at regular intervals, these will provide the soil with the root parts which are left behind in the ground and nourish micro-organisms (see the chapter on Crop Rotation and Green Manuring, pages 68–79).

Adding powdered minerals

Due to the many micronutrients these contain, they serve as a special means to improve soil. Additionally they often act as catalyst and enhancer of all soil life (see the chapter on Compost, pages 52–65.)

Phacelia is an important green manure plant. As pre-sown, after-sown or interim crop it makes a special contribution to soil health.

Garden gold: if we prepare compost properly, after maturing it emanates a pleasant aroma and ensures improvement of soil quality.

Hedge-laying

One of the first things you can do to improve your garden's situation and achieve a good microclimate is to lay a hedge — if landscape circumstances and neighbour relations allow this. Hedges not only protect you and your garden from wind, but offer important habitat and protection to a great number of living creatures such as birds, insects, hedgehogs, frogs, toads etc. (see also pages 105–6).

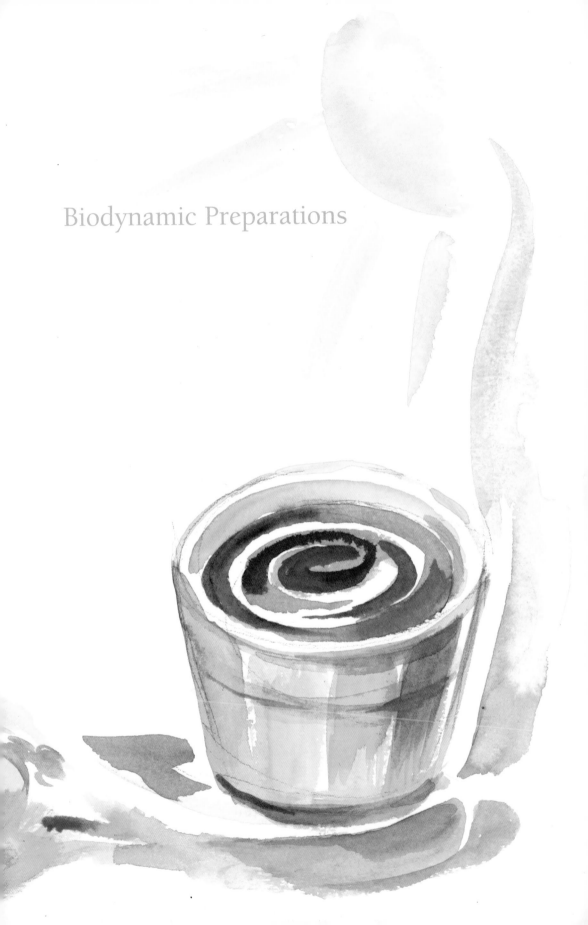

Biodynamic Preparations

Horn manure is one of the preparations which Rudolf Steiner introduced as an important aid for enlivening the soil. It is applied in liquid form.

Of course this isn't true of a domestic garden. Anyone who feels motivated, though, can participate in making the preparations by asking a nearby Demeter farm when it is going to do this, and getting involved. By offering such help you can ensure you procure the preparations you need for your own garden.

Although you cannot usually make the preparations yourself, I will briefly describe the process to help deepen understanding of them. The preparations are, preferably, made on a blossom or fruit day.

Preparations are 'homoeopathic remedies' for the soil, the compost and plants, and are indispensable in practical gardening.

In his Agriculture Course, Rudolf Steiner suggested that the preparations should be prepared on every farm by farmers themselves, and this is still practised widely today on biodynamic farms. The ingredients required are usually available or can be obtained nearby.

Preparations for sale

At the Demeter training and show garden in Binzen, Germany, which adjoins our garden, we make the preparations communally on certain dates in the year, with expert guidance. See page 122 for useful addresses — biodynamic growers and places to purchase the preparations and/or join a Demeter Association gardening group.

The six compost preparations

The five solid preparation ingredients (oak bark, chamomile, stinging nettle, dandelion and yarrow) and the liquid one (valerian) help the compost to rot down well. At the same time they enable the soil to connect better with cosmic forces. To make the preparations we need oak bark, chamomile flowers, stinging nettles before they flower, dandelion blossoms, yarrow blossoms and valerian blossoms. Alongside these plant constituents, Rudolf Steiner also specified animal organs in which the plants are subjected to varying fermentation processes in interplay with the soil and the season.

The plants are collected from a natural environment as far as possible free of pollution. In our garden we cultivate them ourselves, which has the advantage of easy local harvesting at the specified time. It also means we can strengthen these plants with preparation applications, thus enhancing their quality – something we couldn't do in the wild.

Oak bark

For the oak bark preparation we detach loose bark from a mature oak tree (*Quercus robur*). If possible, this is done before midday on a blossom day. The bark is then divided into 50 small pieces and stored until the day we make the preparation.

Dandelion flowers

Every spring we collect the first dandelion flowers (*Taraxacum officinale*). We choose the morning of a sunny flower day and, on their first opening, collect blossoms that have not fully opened yet, without any stem part (the middle of the blossom must not yet have opened). We place them in shade

The oak bark preparation is made from the bark of mature oak trees.

41

to dry, and then store them carefully and undamaged until the preparation is made.

Chamomile flowers

The chamomile flowers (*Matricaria chamomilla*, also commonly known as *Matricaria recutita*) are likewise collected on a flower day morning, shortly before St Johns (24 June) when they are in full bloom but have not yet been pollinated. Pollination means seed formation, and seed-bearing flower heads are unsuitable for making the preparations. We use the flower heads without any stalk part, dry them in the shade and store them in a sealed container until use.

Yarrow

Fully blossoming yarrow (*Achillea millefolium*) are also harvested on a flower day morning: we cut off the topmost blossoms only, with a pair of scissors, and lay them out to dry. Mostly, though, it is more practicable to first harvest more of the plant, with the superfluous stems, and then to completely remove these before making the preparation.

Stinging nettle

The stinging nettle (*Urtica dioica*) is the simplest of the preparation plants to

The blossoms of the yarrow (top) are used to make a preparation. Likewise only the flower heads of chamomile are used (bottom).

process. Whole fresh nettles are mown on a flower day shortly before blossoming (with the first flower sprays already visible), then placed in a trench of 'living earth' of good vitality, and stamped down. Then the trench is filled with soil again.

Valerian flowers

Valerian blossoms (*Valeriana officinalis*) are collected on a flower day morning around St John's tide, with as little stalk as possible. On the same day we place the flowers with water in a shady place to make a cold extract, and allow them to infuse the liquid for a few days, then strain it. We fill dark glass bottles with it and store it until needed in a cool place.

Plants for animal organs

Around Michaelmas, the plant treasures we have collected through the year are made into preparations: we meet with friends of the garden to insert the plants into animal organs. It is now, too, that cow horn manure is made, whereas horn silica preparation (see page 48) is not made until spring.

Using the compost preparations

The six compost preparations are used as a kind of 'vaccine' injected into the compost heap, and a harmonious balance is established there between them. This is best done at regular intervals by

Dandelion (top) must not yet have fully flowered when collected. Valerian flowers (bottom) are applied to the compost in liquid form.

inserting them into the compost container or heap on flower days at descending moon, around every two months. The five solid preparations – oak bark, chamomile, stinging nettle, dandelion and yarrow – are arranged in the heap like the five-pattern on a dice. The distance of each preparation from the next should be at least 20 cm and no more than one metre. The stinging nettle preparation is always placed in the middle, while the other four can be placed in any order. A practical rule of thumb is to arrange them in alphabetical order.

Use as much of the material fermented inside the various animal organs as can be held between three fingers. In other words, these are small 'homoeopathic' doses. They are not scattered but inserted into pre-prepared holes in the compost at a depth of around 40 cm. In the case of long compost piles as on farms or in nurseries, the insertion of preparations is repeated along the top of the heap, at one-metre intervals. The stinging nettle preparation is always placed in the middle of the four others. Finally, 'dynamized' (stirred) valerian blossom extract is finely sprayed over the heap. To do this for your domestic compost heap, add a few drops of valerian blossom extract (stored in dark bottles) to one litre of lukewarm water, and stir in alternating directions for ten minutes. After this, spray it finely, preferably with a garden syringe or pump sprayer, or a hand brush. Then close the compost container's lid, or put some covering over the heap (see page 59).

The five solid compost preparations are arranged in the compost heap like the five-pattern on a dice, with the stinging nettle preparation in the middle.

The horn manure preparation

The horn manure preparation enables the plants to better relate to and connect with their soil surroundings, thus supporting their growth process. The soil is enlivened through the stimulation of numerous metabolic processes that would not take place without its application. This has been very clearly confirmed in the world-renowned DOK long-term trial by the Swiss Research Institute for Organic Agriculture (FiBL). There, for over 30 years, biodynamic (D), organic (O) and conventional (C) cultivation systems have been compared to establish the relative vitality of soil and produce. The trial shows that the nutrient network of the soil responds to the diverse effects of the biodynamic agricultural approach with greater biological activity.

Binding cosmic forces

Cow horns, not the horns of bullocks or bulls, are needed for the horn manure preparation. Cow horns can be recognized by the so-called calf-rings on the horn: one ring forms on the horn for each calf the cow carries. Cow manure is inserted into the horn – it can be collected directly from the fields (preferably on blossom and fruit days). The cowpat needs to be good and firm, for this means it has a low water content, which is important

Cow manure is inserted into cow horns in the autumn. (The horns of a cow, rather than a bullock, are distinguished by its 'calf rings'.)

for the quality of the preparation. Sometimes it is necessary to put one or two cows in the herd on a raw food diet (hay) for a couple of days to achieve the right manure quality. Manure quality in the autumn will be decisive for the quality of the finished preparation, which is dug up out of the earth again in spring.

The filled horn is buried in very enlivened, humus-rich earth. Thus it can absorb the winter's cosmic forces and bind these with the manure. In spring, around Easter, the horn is removed from the earth. The preparation is stored in a stoneware or glass vessel in a box or case in which the vessel is once more surrounded by turf. If properly stored, you can continue to use the horn manure preparation for two to three years.

A market garden of our size (25

A glimpse into the stirring container. Horn manure and horn silica preparations are dynamized by stirring in alternate directions. It is important to form a good vortex.

hectares) needs the material from a large number of horns: through the whole year around 250. The horns themselves can be reused for several years.

Getting the preparation ready for use

The horn manure preparation is now ready to be stirred – or, as it is called, 'dynamized'. To do this you will need a suitable container, preferably a pot or a flagon. It is important for the ves-sel to be cylindrical in shape and have straight sides. It should be made of stoneware, wood or copper. The size of the container will depend on the size of your garden. A garden of 250 square metres, for example, will need a vessel big enough to hold 15 litres. In our market garden we use a large wooden barrel.

I find a suitable stirring place in the garden where I will enjoy spending a whole hour – the length of time that I need to stir it. I use the word 'I' because, in my view, only one person should stir or dynamize the horn manure and horn silica preparation (see pages 48–9) on each occasion, so that a consistent rhythm is estab-lished.

Forming a vortex: the power of depth
I use lukewarm water in the quantity I need for the area I am intending to treat – around five litres of water per 100 square metres, with 25 g horn manure preparation. I stir this quantity in the straight-sided container with a stick or bamboo rod, trying to get as deep a funnel (vortex) as possible to form from outside inwards. To do this I start at the outer wall at a slow speed and increase the rapidity of stirring action as I move towards the centre. Having reached the deepest possible vortex, I pause for the length of two breaths, then place the stick at the outer wall again and repeat the process, but now in the opposite direction and with

The contents of a cow horn. In spring the manure is removed from the horn and can continue to be used for two to three years.

increasing speed. As I do this, 'chaos' and turbulence is first created in the container, and then the liquid starts to heed the laws of my stick so that an ordered rotating movement arises and, again, a deep vortex forms. I stir like this for an hour, as far as possible with the same intensity throughout. You will need to pace yourself. If you also manage, at the same time, to turn your thoughts to the plants which will be growing on the intended area, you are coming very close to what Rudolf Steiner recommended.

Application: large drops for a good effect

After this hour of stirring, the horn manure preparation is applied to the intended area in large drops with, for instance, a hand brush, and the ground is then cultivated for sowing or planting, preferably during 'planting time' (see page 19). This ensures that the preparation does not remain lying on the surface and evaporate. The crop you are going to grow will again determine the right day for this work: root plants on root days, leaf plants on leaf days, blossom plants on blossom days and fruit plants on fruit days.

The dynamized preparation is sprayed in large droplets by simple means, for instance a hand brush.

The horn silica preparation

The horn silica preparation serves plant (foliage) development, and for this reason, unlike horn manure, it is not sprayed on the soil but on the plants themselves in the finest possible spray.

Concentrated sun energy

The preparation is made of translucent quartz crystal with as few impurities as possible. The quartz is ground up into powder, though somewhat coarser grains should also be present (0.5 mm). In the spring, between Easter and Whitsun, the quartz, made up as a paste, is inserted into a cow horn. The horn and its spiral tendency have a concentrating effect on inner, biochemical pro-

Quartz or rock crystal is the basis of the horn silica preparation

cesses. The horn is again buried in very enlivened soil, and absorbs summer solar influences, concentrating these in the preparation. For our market garden the contents of two horns provide the horn silica preparation for use throughout the year.

Stirring horn silica

The horn silica is prepared according to the same basic rules as horn manure (see pages 45–6). Take 0.5 g horn silica preparation to stir into four to five litres of water. As with horn manure, it is best if only one person stirs for an hour. And again, if you can think about the plants in your garden as you stir, your work will take effect at a further, deeper level.

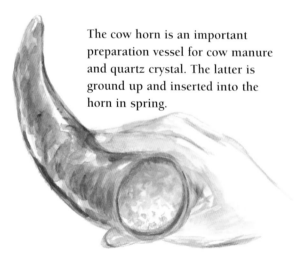

The cow horn is an important preparation vessel for cow manure and quartz crystal. The latter is ground up and inserted into the horn in spring.

Application: the fine-spray effect

After stirring, the preparation is sprayed in the finest form on the whole plants. This is best done with a garden syringe or pump with a fine nozzle.

Start spraying the preparation when the young plants such as lettuce have formed their first four to five typical leaves. Do this in the early morning, and spray on days appropriate to the particular crop, i.e. root plants on root days etc. It is best to start stirring the lukewarm water with horn silica added about half an hour before dawn.

As soon as the plants are more established, and start fruiting or, in the case of leaf plants, are shortly to be harvested, I no longer stir the horn silica preparation in the early morning but towards evening. This supports ripening and aroma formation in the crop.

This quantity is stirred with around 60 litres of water and is enough for an area of one hectare.

Quick-ripening vegetables such as radishes or lettuce are thus sprayed twice: first when the first four or five characteristic leaves have formed, and secondly around three weeks before harvest. For plants that take longer to be ready, such as strawberries or tomatoes, spraying is done five to eight times during the whole growth period.

For crops that will later be stored over winter, it is good to spray on three successive evenings about four weeks before harvest. Choose days assigned to the particular crop. For instance, spray carrots or beetroot on three successive root day evenings, and cabbage on three successive leaf day evenings.

You'll be thinking that this is a lot of work. Yes. But only in this way will your food achieve the best possible life-sustaining quality.

Community stirring

Those who wish to share the 'preparations work', for instance in community garden projects, will do well to find like-minded people and form a 'stirring community', taking it in turns to stir on successive days. Gardening biodynamically can connect and unite people.

Compost – Food for the Soil

The magic formula for compost

Preparing compost means preparing food for the life of the soil, and 'cooking' this food with great love and devotion! The magic formula for the best compost is: Cut up small, mix, moisten and keep covered. Observing these rules is more important for successful compost than making it at the most favourable constellation. It would be good if the outstanding importance of compost for the health of plants was recognized in our daily language. Instead of talking in a derogatory way about composted 'waste products', for instance, we should call them something like 'composting ingredients', since the high-quality compost they can produce gives vital nutrients to the soil.

A good tool like these secateurs makes compost preparation easier

Cutting up small

Rule one is to cut up the ingredients. The most important thing when making compost is to provide 'mouth-sized' morsels for the micro-organisms that are involved in diverse ways in breaking down and converting substance into humus. All organic ingredients should therefore be cut up small – preferably into thin strips or small cubes measuring four square millimetres. The cut surfaces offer purchase for micro-organisms that will quickly break down the organic remains.

All remains of food can be added to the compost heap, including sausage, and lemon and banana peels. The only proviso is that they must be cut up small so that micro-organisms can dine easily on them.

Everything (can go) on the compost heap

All remains of food can be added to the compost heap – not just fruit and vegetables (including cooked vegetables) but also bread, cheese, pasta, sausage, etc. Citrus and banana peels (preferably from organically grown produce) can also be composted. Try the following yourself. Place a halved, squeezed lemon in the compost as it is, and cut up the other half into the recommended size (4 × 4 mm). See what happens to both within a week. You'll be astonished!

Larger cut materials will not achieve optimum conditions, since the micro-organisms will need longer to break them down; and undesired side effects will also arise.

Valuable organic materials

Eggshells can be put in the compost. They are an important natural source of calcium. The membrane adhering inside is also a tasty feast for the numerous living organisms in the compost heap. But eggshells, too, must be cut as small as possible. The best thing is to crush them in your hand and place them straight into the kitchen compost bucket. If you prefer not to do this, collect the eggshells in a bag and crush them later with a rolling pin.

Preparation is everything. There's really nothing that can't be added to the compost heap. The important thing is to cut up the organic ingredients very well. You can do the major part of this with a spade.

Avoiding unpleasant smells

Our family collects organic food remains in a separate compost bucket. Make sure your container isn't too big, for it ought to be emptied at least once a week.

It is worth already adding rock dust (see page 58) in the kitchen to organic remains, to avoid rotting. For this purpose you can convert a glass container (approx 500 ml) with screw top into a giant salt shaker. Punch about ten holes in the lid and fill the container with the best rock powder that you can buy. Then, whenever you put more organic remains in the compost bucket,

A Swiss 'Gertel' is the best tool for cutting up softer, less woody organic ingredients. It has a straight blade and horizontal handle.

sprinkle them with a little rock dust. This will not only support the digestive processes of micro-organisms but at the same time prevent unpleasant smells around your bucket.

Siting your compost heap

The shape of your garden compost container is only of secondary importance for the success of the compost. The most important and, in fact, indispensable thing, is to chop materials up small. A shady spot is good for the compost container since in direct sunlight the compost will dry out too much at the edges. The con-

Cuttings from trees or bushes are especially suited as chopped materials for the compost heap, and likewise grass mowings. However, the plant material must be very well shredded or cut up before being added to the compost, so that micro-organisms can make a meal of it.

The Swiss Gertel being used for chopping up organic compost ingredients

ment and where necessary intervene to correct it.

The compost heap: filling and mixing

1. Fill the compost container to about 20 cm with a first layer of coarse twigs so that a drainage layer is formed, allowing moisture to seep

If you decide against a compost heap in the garden, you can buy a range of compost containers instead, in varying shapes. Make sure they are open at the bottom, so that soil organisms can migrate inside them.

tainer should also be open at the bottom (although this does not of course apply to so-called 'balcony compost'). You can take pleasure every time you pass by in seeing the processes at work in it, so it does not need to be hidden away at the bottom of the garden.

Remember that you're a 'top compost chef' preparing the best nutrition for soil organisms. The more care you put into preparing it, the better you will achieve, within six months, a transformation from organic substances into the best, crumbly humus, smelling of forest earth. For this process to occur, you need to accompany its develop-

away where necessary into the ground.

2. Get hold of shredded bush or tree prunings or make these yourself. You should always have sufficient quantities of this material available. 'Shredded' here means the fibres are opened lengthways, for instance with a hammer shredder rather than a disc-wheel shredder with a smooth knife cut – which does not offer purchase to micro-organisms.

Whenever you put organic remains in the compost container, add these shredded clippings too, as a third of the new amount. This also applies to composting grass mowings, which must likewise be well

The valerian preparation as compost application is dynamized for ten minutes in a container (e.g. of earthenware).

mixed with clippings since otherwise they will stick together and putrefy.

3. As soon as new material is added to the compost, mix this thoroughly with the top (last) layer – to the depth of a fork, or about 30 cm. This works best with a handy, five-tined farmyard fork, as the gap between the tines is right for easily working with mixed prunings and organic remains. Regular mixing is important for avoiding distinct strata in the compost, and ensuring the materials rot down well. It avoids putrefaction and enables the micro-organisms to work speedily enough without losing valuable nutrients in the process.

4. Keep the compost moist. Whenever you add new, mixed materials to the pile, and have mixed them in as recommended, test the current

This five-tined farmyard fork is excellent for compost work: both for mixing and filling the container with organic ingredients.

56

moisture content of the compost with the so-called fist test. This involves first wetting the material that you have just added with a watering can with watering rose — the water should be as finely distributed as possible. Then take a handful of this mixed material and make a fist around it. No water should seep out, and when you open your hand again the material should continue to stick together. Push the clump with a finger, and it should fall apart lightly. If this happens, you have achieved the right degree of moisture. This simple test is something you should do every time you come to the compost heap, even when you are not adding new

material, and preferably every day. The more often you mix the materials, the more you enhance the activity involved in transforming organic remains into the most fertile humus.

5. If water seeps out in the fist test, the compost is too moist and more shredded prunings must be added, which will absorb the excess moisture. If the compost is too dry, and the clump therefore falls apart when you open your hand, adequate water is sprayed on finely and everything is mixed again. Mixing in clippings or prunings is vitally important for air circulation in the compost so as to prevent anaerobic processes of putrefaction. You should therefore keep adding fresh material to your compost container

To keep the compost moist, use a watering can or hose with watering rose, since the water should be finely distributed.

throughout the year, even in winter. This is the only way to maintain the conversion process involved in decomposition.

6. Keep the compost covered so that it retains optimum moisture content. Avoid direct sunlight on the rotting material. In other words, always keep the lid on a compost container or put a covering material such as fleece, coarse sacking or old carpet on an open heap.

7. If you really adhere to the golden rules of 'chopping – mixing – keeping moist and covered', there is a very good chance that you will succeed in creating a top-quality compost. Then the soil organisms will rejoice with burgeoning fertility. Sustained compost quality will be achieved by repeatedly adding the biodynamic compost preparations to the active rotting zone at two-month intervals, in the order described. Happy composting!

Use of volcanic rock dust

In our garden in Germany we use Vulkamin volcanic rock dust, both in our compost and soil preparation, because of its range of outstanding properties. (In the UK, SEER Rockdust, which is also very rich in minerals and trace elements, is obtainable in 20 kg bags, see Useful addresses, page 122.)

The rock dust is scattered on the soil surface and lightly raked in. Earthworms digest it, producing mineral-rich casts. With increased mineral availability and microbial activity, soil fertility is much improved. Among the benefits are higher yields, crops have a greater resistance to pests, the nutritional value of fruit and vegetables is higher, their flavour is better and shelf life longer.

When making compost, sprinkle rock dust over each fresh addition of material to the heap.

The heat of your compost can be checked using a compost thermometer. The temperature should be between 45 and 70°C.

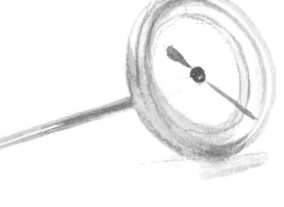

The golden rules for composting

1. Chopping
2. Mixing
3. Keeping moist
4. Covering

Fleece for an open heap

Everything that has so far been said about composting in a compost container also applies to the classic compost heap or windrow. But here it is essential to cover the heap with compost fleece.

'TopTex compost fleece' is a breathable fleece made of 100% stabilized polypropylene fibres. Its properties and formats are in line with diverse agricultural requirements. It is an important aid to optimum, controlled composting and simple to use. Its biggest advantage is that the fleece maintains optimum moisture conditions in the windrow both by diverting precipitation from the surface and protecting against sun and wind. In addition, it ensures aerobic rotting and optimum temperatures, even at the margins of the heap. It allows the necessary gaseous exchange to occur and minimizes seepage and loss of nutrients. The golden rule of 'covering' can thus equally be applied to open heaps, ensuring top compost quality.

If you decide to have an open compost heap, the golden rule of covering the heap still applies, which you can do by using fleece, coarse sacking or old carpet, to maintain optimum moisture.

Maturing compost

After 15 to 30 weeks the compost is ready and can be used in whatever way your plants and crops require.

Young compost

Depending on intended use, i.e. for potatoes, cabbage, leeks and celeriac as well as other heavy-feeding vegetables,

you can apply a well-rotted compost that has not yet fully turned to earth after about 15 weeks.

For more delicate vegetables (light feeders) such as carrots and parsley, as well as for growing on young plants and to provide sowing soil, you will need a thoroughly rotted compost that is fine and crumbly and has a forest soil aroma. Take care if it smells of sulphur: then the compost has not finished rotting down, and sulphuric water prevents speedy growth of plants that come into contact with it.

Little quality test

A good, simple test of compost quality can be carried out at follows. Take a well-mixed sample of your compost, place it in a seed tray and sow cress on it. After watering, put the tray in a plastic bag so that moisture is retained. Let the cress germinate and grow to a height of about 10 cm. You can tell the quality of your compost by the happy growth of the cress. If it does not grow properly or the roots are brown, the compost is not yet mature enough and the rotting process has not finished. In this case the compost should be well mixed again and 'vaccinated' with the

For good growth, light feeders such as carrots and onions need an average of five litres of compost per square metre

biodynamic preparations. After another four weeks it will be ready to use.

First test the compost quality, therefore, by using all your senses, and carry out this cress test before using the compost. This will prevent unpleasant surprises.

Heavy feeding crops like cabbage, cauliflower and leeks need on average 20 litres of compost for healthy growth.

Medium feeders such as kohl rabi and Swiss chard are given about twelve litres of compost per square metre while they are growing.

Mature compost

Lovingly prepared, mature compost is applied as soon as possible to the soil. The longer it stays unused the more its efficacy wanes. The best time to use it therefore is from mid-March to the end of October.

Celery is a vegetable that has moderate nutrient requirements.

Vegetables like carrots and beetroot (top right) are light feeders, and need well-rotted, mature compost

During this time, due to the warmth of the soil, it will be best able to exert the desired 'fermentation effect' in the soil. Soil warmth facilitates good microbial activity in the ground, which, no less than compost making, is a prerequisite for optimum biological soil activity.

In the period mentioned, compost is applied together with a general soil preparation (horn manure). The compost is worked into the soil to a depth of 10 cm. In general, a root day is always best for this, unless you have decided on special soil preparation to strengthen the particular type of plant you will be growing there. Then for cabbages, for instance, you can apply the compost on a leaf day.

Please don't leave a single crumb of your lovingly prepared compost lying unwanted on the surface. In only a few hours the sunlight will destroy its diverse, valuable constituents. In addition, the next rainfall will wash out important active ingredients. It is therefore definitely advisable to work the compost well into the soil rather than applying it to the surface. Avoid applying compost in late autumn and winter: it is much too valuable to waste in this way.

Fertilizer is for making fertile

And therefore we should be very sparing when using this type of cultivation aid.

Commercial fertilizers

Make sure that you do not buy artificial fertilizers but only ones that are authentically organic – read the small print on the bag. There are classic fertilizers made of animal substances such as horn and bones, and also purely plant-based ones made, for example, from shredded castor-oil plant or nitrogen-rich leguminous plants such as field beans.

The biggest group of fertilizers com-mercially available are so-called mixes, containing animal excrement – thus various forms of manure. They often come from overseas, e.g. guano, which is sea-bird droppings. Or they come from intensive rearing, e.g. chicken manure. You should really only use these products if you haven't made enough compost.

Only ever use commercial fertilizers if your own efforts have not been so successful. And this shouldn't of course stop us from seeking continually to improve our own gardening practice. And no opportunity should be missed to add a little well-chopped material to the compost heap, and mix this in well. Buying in commercial fertilizers can help us in the short-term to address a lack of nutrients if our soil analysis shows this is needed.

The main shortage is likely to be nitrogen, since in the domestic garden

You can tell from the way plants develop whether they have enough of the right nutrients. Left: well-fed cabbage. Right: poorly nourished cabbage

63

we rarely have much opportunity to work really good cow manure into the compost. We need nitrogen especially for our heavy feeders such as leeks, celeriac, all types of cabbage and potatoes.

The seeds of the castor-oil plant can be shredded to make a fast-working, organic, commercially available fertilizer.

Two kinds of application

1. Add the fertilizer with the soil preparation to the soil and work it lightly into the surface. Depending on the fertilizer and its density, it will take from two to four weeks for the fertilizer to be transformed by soil organisms into nitrogen that is available to plants in the form of nitrate. With this method, therefore, it is important to know exactly how quickly this occurs with each type of fertilizer. A rule of thumb is that the finer it is, the more quickly it will be activated, and the coarser, the longer-lasting will be the flow of extra nutrients in the soil. Only in emergencies should one use so-called multi-nutrient fertilizers. Usually it is nitrogen that is lacking. Phosphate and potash are often available in the domestic garden in excess quantities. Make sure that the prime flow of fertilizer falls in the main growing period of the crop, and that, particularly at the ripening phase when vegetative growth is concluding, plants are not 'driven on' too much with excess nitrogen. This will severely impair the storage quality of crops and their taste. With too much vegetative impetus, too, the subtle impulses that need to be stimulated by drawing on the constellations (the moon and stars) cannot unfold properly in the plants.

2. Make a so-called 'heavy-feeder mix' from the commercial fertilizer and a good compost. Use about 75% compost and 25% fertilizer, mixing them together to make a special compost heap early in the spring, so as to achieve quicker biological activity with the commercial fertilizer. This application will not have such a strong 'driving' effect on plants.

The harmonizing process can also be positively accompanied by making the right use of the compost preparations. Here Rudolf Steiner's statement that 'fertilizer must become more sensitive' really comes into its own.

You should allow around four weeks before working the specialized compost into the upper surface of the soil, at the right constellation and supported by the horn manure preparation.

Since heavy-feeding crops stand in the beds for a long time, it can be advantageous to lightly work in some of the specialized compost again between the rows about two months after planting, and then cover the spaces between with a good layer of mulch.

Our top priority should always be to actively nurture the soil, and make the right use of green manures. A living soil will offer sufficient nutrients for plants if we take care to plan a balanced crop rotation.

As a short-term measure, nitrogen can be applied in the form of commercial (organic) fertilizers. Leguminous plants collect nitrogen from the air on a longer-term and more sustained basis, and their roots penetrate the soil much more extensively than leeks.

65

Crop Rotation and Green Manuring

What does crop rotation involve?

Crop rotation is an important tool for the organic gardener, supporting the health of plants and good resistance to disease. It involves documenting the succession of crops in one's garden over many years, so that plants belonging to different plant families are cultivated in alternation through a season or several years.

This alternation, or rotation, is important for ensuring that the soil does not become exhausted. Members of the same plant family add similar plant secretions to the soil and take similar nutrients from it. Targeted alternation of plant families therefore makes it possible to maintain the diversity of nutrients in the network of the soil.

A good plan for your beds

The period soil takes to recover from cultivation of any type of plant depends on the particular plant family. On pages 70–3 therefore, the most important types of vegetable are assigned to their respective plant families, along with the corresponding 'recovery' time for the soil.

You can draw up your own crop rotation plan based on these periods. In other words, plan ahead for each year which crops you are going to grow and where. To do this, use a 'garden plan' – that is, a drawing which sketches the separate beds in your garden, where you write down the crops and their positions. This documentation serves as the basis for planning cultivation in successive years, and helps to avoid errors of rotation.

In your garden calendar, check plantings of recent years in each bed. Cruciferous plants can only be cultivated in the same place again after a certain interval.

Green manuring: helping the soil regenerate

Green manuring is an important soil-enlivening tool in crop rotation. With their complex root systems, green manure plants leave behind a great many cavities in the soil that loosen it and enrich it with nitrogen. Important green manures are: field beans, field peas, Persian clover, red clover, white clover, berseem clover or Egyptian clover, sweet clover, lupin and lucerne or alfalfa.

Lucerne is also called the 'queen of green manures'. Its roots can delve as far as 15 metres (!) into the soil, thus opening and loosening it to a great depth. It transports nutrients from deeper layers of the ground to the surface. With a cultivation period of three to five years, lucerne is one of the longer-term green manure plants.

Red clover is an important green manure plant. With a root up to two metres in length, it makes a considerable contribution to soil fertility.

For an extensive root manuring effect, lucerne needs to be in place for three to five years, and so you should take account of this in your crop rotation plan.

Recognizing and understanding plant families

Crucifer or cabbage family (Cruciferae, also called Brassicaceae)

The family of cruciferous plants includes all cabbages, Asian vegetables, Chinese cabbage, broccoli, turnips, kohl rabi, swede, cress, radish, rocket, etc.

Crucifers drain the soil of nutrients very quickly, and therefore this family requires the greatest interval between crops. A bed where cruciferous plants have been cultivated should not be planted with the same family again for four years. In other words, in the fifth year you can grow them on the same bed again.

Nightshade family (Solanaceae)

This family includes tomatoes, aubergines, pepper (capsicum), Cape gooseberries and potatoes. The five-petalled flowers often grow together to form a tube or funnel. Members of this family can, if necessary, cope with another crop in the following year on the same ground. However, it is better for the soil to leave an interval of two years.

Goosefoot family (Chenopodiaceae)

This family includes spinach, beetroot, spinach beet, garden orache, Good King Henry, Swiss chard and quinoa.

The plants have small, inconspicuous flowers, which often grow together to form clusters.

Goosefoot plants, like the Cruciferae, are sensitive to direct replanting, and a

pause of at least three years is necessary. In addition, they do not like subsoil waterlogging or compacted soil, where they will fail to grow properly. Spinach grows yellow and beetroot fails to thrive.

Daisy family (Asteraceae, also called Compositae)

This family includes lettuce, endive, Italian chicory, chicory, radicchio, scorzonera, Jerusalem artichoke, cardoon, dandelion, pot marigold, etc.

Each 'flower' consists of tiny individual florets which are packed closely into a compound flowerhead, like a 'basket'. Plants of this family can happily grow on the same plot after just a one-year interval.

Lily and onion family (Liliaceae)

The lily family includes all types of onion, thus Welsh onion, spring onion and bunching onion as well as leeks, garlic, chives, lilies, tulips, hyacinths, etc.

The plants of this family form a bulb as 'organ of survival'. Their

The lovely sun-flower belongs to the family of Asteraceae (Compositae). Their large flowers form a basket-like array of florets, which is an easy way to recognize them.

leaves have parallel veins and no indentations.

An interval of at least two years is needed for these plants.

Umbellifer family (Umbelliferae, also called Apiaceae)

This family includes carrots, parsley, celeriac and celery, parsnips, dill, coriander, etc.

The flat-topped flowerhead characteristic of this family is called an

If you let carrots grow flowers, they will form typical umbels, and therefore they belong to the Umbelliferae.

Soil nutrient network

Through intervening sowings of green manure plants, the garden soil is given new impetus for its fertility. The positive effect of green manure plants has been confirmed not only through decades of gardening practice but also by recent research. The latter has shown that the soil's nutrient network is a far more complex system than was previously assumed in conventional views (DOK trial).

The diverse interactions in the soil's nutrient network – for which only initial research so far exists – lead to a continual mutual giving and absorbing between the plant and its environment. Through its root secretions every plant nurtures its own distinct microbial milieu in the direct vicinity of the roots and, with the help of this, soil life in turn provides plants with the complex of nutrients that they need.

umbel, and looks rather like an umbrella.

Umbellifers need at least two years between crops on the same plot.

Pea and bean family (Leguminosae, also called Fabaceae)

These include runner beans, bush beans, peas, sweet peas and vetches, lupins, clovers, etc.

The distinctive individual flowers of this family are composed of a broad upper petal (the 'standard'), two side petals ('wings') and two lower ones that together look like a little boat (the 'keel').

The interval between crops on the same ground is at least three years, and for peas at least five years.

Leguminous plants are very good as preparation for subsequent crops since they add nitrogen to the soil. You can make use of this fact by intentionally leaving their roots in the ground when preparing the plot for the next sowing or planting. The roots of leguminous plants live in symbiosis with special bacteria, of the genus *Rhizobium*. These have the capacity to bind atmospheric nitrogen and transform it into soluble nitrogen compounds that plants can use. These nitrogen nodules are then available to following plants as nutri-

Vetches also belong to the Leguminosae family. Plants in this family have typical pealike flowers in which the five petals are fused together to form the 'standard', 'wings' and 'keel'.

tion within the soil's biological network.

The capacity of leguminous plants to add nitrogen to the soil can be used still more intensively in your garden by cultivating them specifically as green manure plants.

Green manuring in practice

Whenever you sow green manure plants, prepare the soil in advance on a root day and also spray horn manure preparation (see page 45). Then the seeds are sown, either deeper or more shallowly depending on type of plant and size of seed. While peas and beans, for example, are sown in drills about 4 to 6 cm deep, fine clover seeds are placed in the top 0.5 to 1.5 cm of soil.

Green manure plants are usually sown somewhat closer to each other than ordinary vegetables to achieve quicker ground coverage and to suppress weeds.

The growth of green manure plants is supported during the so-called four-leaf period (when the plant has developed four of its typical leaves) with a morning horn silica spraying on a root day.

Short-term green manuring

For quick green manuring of between four and eight weeks in the spring, field beans are very good.

If you wish to plant leeks at the end of May, for example, field beans are planted prior to this in February/March. Plant them thickly, every 2 cm, and at a depth of about 4 cm. The distance between drills is narrow, about 10 cm, and at the end of April the parts of the beans above ground are

removed. The fresh leaves and stalks are a welcome addition to the compost heap, since the plant sap of a fresh bean plant gives good moisture. Thus only the bean's roots stay in the earth.

For a quick green manure in summer, likewise lasting between four and

If one intends to plant root vegetables, a 'roc manuring' should be carried out beforehand to loosen the soil well. Carrots, especially, w otherwise tend to go leggy

eight weeks, berseem clover is particularly useful. It is quick growing, an annual, and cannot survive the winter. In general, clovers improve the soil a good deal. The fine network of roots that they leave behind, with the atmospheric nitrogen they fix, makes an excellent tilth.

Longer-term green manuring

Longer-term green manures grow for between four months and a year. While this may seem like a long time, it is highly recommended since the positive effects on the soil's nutrient network are considerable.

White or red clover is suitable for green manuring over this longer period, as both types survive over winter.

Once again, soil preparation before sowing is done on a root day. After about four weeks, on a root day morning, the plants are stimulated with a spraying of horn silica preparation.

After a further two weeks, the plants are mown on a root day, shortly before they blossom. Mowing should not cut the plants off at ground level but leave about 15 cm standing. Add the chopped-up mowings to the compost heap. By cutting in this way the plants' root growth is stimulated once again, since cutting the upper parts and thus reducing leaf mass means that around 30 per cent of the roots die back. These dying roots are digested by the soil organisms. After a percentage of roots have died, the plants once again form many fine root fibres and put forth new shoots above ground.

A short-term green manure using field beans is good preparation for leeks

By applying horn manure preparation to the mown area on a root day evening you additionally support the soil's nutrient network in a special way. The evening is a very good time of day for this because, as the earth breathes in in the evening, the preparation can penetrate the earth better and unfold its beneficial effect.

Root manuring in winter

You will now no doubt understand why instead of, or as well as, using the term 'green manure' I also speak of 'root manuring'. One really cannot overestimate the value of cultivating deep-rooting green manure plants that

This young vetch (a leguminous plant) already displays root nodules containing bacteria that bind atmospheric nitrogen, making it available to other plants.

have enough time to grow at a particular site.

Proven mixture: rye with vetch
Particularly in the cold season, also, root manuring is needed. By cultivating plants on winter beds we can impede the leaching of nutrients. Sow an over-winter root manure, therefore, on all garden beds that are no longer needed for overwintering crops. The following mix has proven its worth over many years: 80 per cent winter rye and 20 per cent winter vetch. Rye is a plant with a well-developed, fine network of roots and is therefore excellent for loosening the ground. Did you know that a single rye plant can form up to 600 km of root? The vetch helps in collecting and fixing atmospheric nitrogen.

I sow this rye-vetch mix in my market garden on almost all beds, always in the autumn after harvest, up to the beginning of December. Depending on weather conditions, the plants germinate either in December or January.

The root manure mix is not just beneficial for the soil but also pleasing to the eye. It is a lovely sight in winter to see the red glimmer of germinating rye plants. The first sunny days at the end of winter soon help the crop to cover the ground.

Mulching: giving the beds a good blanket
However, a few beds and areas are not sown with this root manure. These are

the ones where we will first grow carrots, lettuce or onions in the spring. Cultivating these crops requires soil ready to plant at such an early stage in the year that a green manure will not have its full effect. Instead, in the autumn, we treat these beds with a deep loosening of the soil (without turning it over). In early spring we undertake a shallow working of the soil, giving the appropriate cosmic impetus to the plants that will be grown there (root vegetables on a root day etc.) supported by application of horn manure preparation. This shallow working of the soil allows the first weeds to germinate, and the surface of the soil also quickly dries.

Over winter it is advantageous to give a mulch cover (see pages 94–5) to beds that have not been planted. This is removed in spring before planting. Do not work the mulch cover into the soil since this will not provide good conditions for germination.

Sowing flowers to help renew the soil
If you chiefly grow vegetables and other food crops in your garden or fields, you are unlikely to have many flowers. To give the soil an additional fruiting impulse, therefore, we regularly sow flowering plants. If we consider the plant in terms of its overall development, we can see from our work with the constellations that the blossom is one of the four developmental characteristics, all of which

Phacelia has many benefits: as bee pasture, promoting plants' fruiting, and as excellent root manure plant.

should be nurtured. It is therefore a good idea when planning your crop rotation to incorporate the blossoming impulse (see also plant protection, p. 100).

For instance you can regularly transform your vegetable beds into colourful flower beds with a rich mix (but make sure you include none of the composites). Or you can sow individual types of blossoming plants such as borage, phacelia, pot marigolds, cornflowers, nasturtiums, etc. Many flowering plants can also be used in the kitchen.

For soil preparation, choose a root day again (and other preparations as indicated previously).

When the burgeoning blossoms unfold a few weeks later, spray the horn silica preparation on a root day evening so that, as the flower fades, the plant's strength can work back more into the root. If you want to prevent the flowers seeding themselves on the same bed the next year and competing with vegetables, pick off the ripe seeds in good time and keep them for another sowing next year. Or mow them in the early morning in two stages. The early morning is important since there are few insects about then that might be harmed. Two-stage mowing means that on one day you first mow a part of the bed or area, and two days later the rest. If you only have one flower bed available, only half of it is cut. When this mown area starts blossoming again, cut the second half. Thus insects have enough time to look around for more flowering plants. The cut vegetation is chopped up and added to the compost. Treat the 'stubble' on a root day evening with horn manure preparation again.

Example of a six-year crop rotation

The following gives an example of a six-year crop rotation for a bed. Use your garden plan (see pages 68 and 82) to draw up and document your crop rotation. Your planning will be based on knowledge of the plant families and the resulting intervals between crops, as needed by the soil's nutrient network.

It makes most sense always to start and end a rotation cycle with a root manure crop. Soil treatment with the horn manure preparation, and spraying of horn silica, is of course done every year, and is therefore not mentioned additionally below.

Year One

In the first year after root manuring, one can start with potatoes, which are heavy feeders. They need a nitrogen-rich soil that has previously been provided by the root manure. At the same time as planting on a root day, a large quantity of good compost is applied to the surface.

Instead of potatoes, it would also be possible to plant tomatoes, aubergines, pepper and other fruit plants, since all such heavy feeders will appreciate the well-fertilized, 'rested' soil. In the case of fruit plants of course, cultivation work and horn manure applications should be done on fruit days.

In summer, after the potato crop, spinach or lamb's lettuce can follow. For this you need only work the soil lightly and superficially, on a leaf day.

Year Two

On this bed, leeks or cabbage can now be planted (leaf plants on leaf days, with the exception of broccoli, which prefers blossom days) and a good amount of compost given. It does not matter whether the cabbage is already harvested in autumn or, like Brussels sprouts, stands over winter.

At the end of the second year, the

cabbage crop can be followed by an overwintering sowing of the winter rye and winter vetch mix as root manure.

Year Three

Root vegetables can now be sown on root days on this bed, e.g. carrots, beetroots, turnips, swedes, parsnips or Hamburg root parsley. If you planted a root manure the previous year, this must be removed from the bed fairly early on (depending on weather conditions, at the beginning of March up to the middle of that month). Do not work any greenery into the soil, since root vegetable seedlings do not like decomposition processes in the soil.

Year Four

In spring the soil of this bed is strengthened with mature compost, and then alternating crops of light feeders such as lettuce or other quick vegetable crops like radish and cress are cultivated. The constellation when preparations are applied is oriented directly to the particular crop – that is,

In the third year of the crop rotation, a root vegetable such as beetroot can be planted. This is sown on root days during planting time (see p. 19)

lettuce on leaf days, radishes on root days, etc.

Year Five

Soil fertility is supported during this year by blossom plants or herbs such as basil, dill, borage, etc.

Year Six

It is now time for a full root manure treatment, preferably for a whole year, so that the soil's nutrient network can communicate generously with the plants via their roots. White or red clover is suitable for this. The clover is repeatedly mown and can be used as mulch or chopped up and put on the compost heap.

Cultivation Work and Care of the Garden

Planning and preparing

The gardening year usually begins in autumn and winter, when the new seed catalogues come through the letterbox.

The garden calendar: planning for the new gardening season

This is when the past year's garden calendar really comes into its own (crop rotation, see page 68). Here, hopefully, you will have recorded garden 'occurrences', such as notes about the success or failure of various crops. Such information can help you decide what to do in the coming year. For instance, you may have noted that your onions didn't ripen properly because there wasn't enough sunshine where they were sited. Or you might decide not to plant tomatoes again, since they have continually been affected by blight. However you could apply an equisetum (horsetail) preparation to them (see page 107), and see if this improves matters. Of such things, too, of course, you can make notes. It is also important to note which varieties were cultivated when, on which garden beds (crop rotation, see pages 68–79).

In our market garden in Binzen, Germany, we draw up the cultivation plan each November for the following year. In an initial overview we ascertain the sowing and planting dates for each type of vegetable within a period of ten days or so. Fine planning for the right constellation will then be done in January.

This general planning is best done in the form of a grid on which the individual months are marked at the top and the separate beds down the side.

The decisions made here are transferred to the work diary for the current year, in line with the *Biodynamic Sowing and Planting Calendar* (see page 123), which can be ordered from bookshops.

No time for your garden?

If you can't tend your garden for a while, perhaps because you're away on holiday, you should include this in your planning. Likewise, if there are usually several consumers of your self-grown vegetables, you also need to consider whether there will be times when some of them are away — and try to ensure that only smaller quantities need harvesting then. Of course this depends on knowing in advance how long, say, radishes, lettuce or other types of vegetable need to mature in different seasons. Here once again, the observations you recorded the previous year will be an invaluable help.

This shows the constellation con-figurations for the current gardening year with indications of the best sowing and planting times. Thus, before the year begins, you will know when you have to do certain jobs — as long as the weather gods allow!

'Planting well is half the trick' as an old phrase has it. And to those who say that they can't make plans ahead of time because of all the unpredictable factors, I would reply that good plan-ning, no matter how the weather plays up, is a thread to follow through the year, something you can deviate from but always come back to again. Thirty years' of experience in our market garden have taught us this.

Soil preparation

This preparatory work provides a good basis for subsequent plant growth. Choose a time for this work that accords with a favourable con-stellation for the type of crop (root, leaf, blossom, fruit) you are cultivat-ing, as long as weather and soil condi-tion allow.

Thorough soil cultivation

This is done as follows. First dynamize the horn manure preparation (see p. 45). After an hour's stirring, apply the preparation immediately to the bed in large drop-lets. Then loosen the soil (without turning it over) with the aid of a gardening fork. If you are going to grow heavy feeders such as cauliflower, celery or leeks, it is good to apply a large quantity of well-rotted compost, which is worked in to the top 5–10 cm of the soil. If you do not (yet) have any compost available, you could also use an organic commercial fertilizer. Good compost, vaccinated with the biodynamic preparations, would however be the first choice since it nourishes the soil's nutrient network in a comprehensive way, and stimulates more life and vitality.

For root vegetables, especially carrots, parsnips or root chicory, soil cultivation is carried out to as great a depth as possible without turning the soil. Leggy roots in these vegetables are always a sign that the soil is too compressed.

83

Sowing out in the open

For sowing out in the garden, the ground is prepared as described on page 83. Most vegetable seeds are sown at a depth of around 0.5 to 3 cm. Where possible, the seeds are placed individually into the seed drill at the correct intervals. A piece of postcard-sized paper is a practical aid for this. Fold it in half, place the seeds in the fold, and use a knife to detach one at a time at the right interval. If you manage to draw the seed drill as straight as possible on the bed, hoeing later on will be much easier. Stretching a string as guideline makes this easier.

Afterwards, mark the row with a weather-resistant label on which you record sowing date, type of vegetable and variety. Then you won't forget what you sowed where. It is then pos-

Keeping the right distance

The most important thing when sowing seeds is to ensure they are at the right distance from each other in the row. This is more important for a successful crop than the distance between rows. The distance between seeds is governed by the later size of the full-grown plants.

sible to identify the germinating plants and distinguish them from undesired seedlings (weeds etc.) which are removed. At the same time this will school your observation so that next year you'll be able to tell plants by their seedlings.

Weed or vegetable seedling? Carrot seedling (left) and chickweed (right) look very similar. However, they can be distinguished fairly easily by the reddish neck (hypocotyledon) of the carrot.

Distance for vegetables

A distance between rows of 25 cm is usually enough for most types of vegetable, apart from cabbages, celery, leeks and other plants that grow big. The larger types of vegetable are planted in rows between 50 and 75 cm apart.

To make sure the seeds are kept in line, a planting string can be helpful. The most important thing is the distance between seeds in the row.

If you only have a small garden, or your beds are too small for such things as Brussels sprouts, courgettes and celeriac, you could plant these singly in certain places. You might for instance place them in between two beds.

Carrots should be set between 1 and 1.5 cm apart, and Hamburg root parsley, parsnips and beetroot between 4 and 5 cm. Spinach and red radish grow very well at intervals of 2.5 to 3.5 cm. Larger white radish need between 15 and 25 cm of distance between each other in the row, and the white Japanese radish as much as 30 to 40 cm. These examples are taken from our practical experience in the market garden.

Too many in a row: thinning out

If the seed is too closely set, thin it out as soon as possible to the intervals specified above.

If the variety allows this, seedlings pulled out when thinning can be replanted elsewhere. Beetroot can easily be replanted, but carrots cannot and will usually grow leggy if you

For your garden calendar

In your garden calendar, alongside the date of sowing, vegetable type and variety, also record the depth of the seed and the interval within a row. Documenting things in this way will enable you to compile useful records that will serve you well in future years.

replant them. Wherever possible try to avoid sowing too thickly, for this will mean unnecessarily thinning too many seedlings – which requires work, and costs money for wasted seeds.

Good seed in good earth. But sowing itself should be carried out carefully so that the seed finds its right way into the soil.

Make certain that old seeds can't germinate! To do so, put them in an oven at 150°C and then mix them with seeds that will germinate in a 3:1 ratio. This is a simple way of avoiding over-density of sowing.

Experience shows that sowing seeds too thickly occurs mainly with carrots and lettuce. You can avoid this by the following means. Mix dry sand with the seeds or really old seeds of the same kind that can't germinate, and then the intervals between seedlings will inevitably be bigger.

Another way to ensure proper gaps is to use commercial products such as seed pills or a seed ribbon. Seeds in this form cannot be stored for long, however, so only buy what you need for one year.

Good thoughts when sowing seeds

As well as checking the intervals between seeds when sowing, you can also bring calm and composure to this activity. Remember that sowing is a kind of 'sacred' act. In former times, sowers would traditionally often speak a sowing verse to accompany the seed on its journey into the soil.

Sower's verse
Measure your step, measure your
  swing!
The earth will keep on being young!
There falls a grain to die and rest.
Its rest is sweet. It lies at peace.
Here's one that pierces through the
  clod.
It finds its goodness. Sweet the light.
And not a grain falls out of this world.
And each falls as it pleases God.
                Conrad Ferdinand Meyer

Helpful tip: weed-free beds

Here, finally, is a useful gardening tip.

If you want to have 'clean', that is weed-free beds – for instance when sowing carrots – I recommend the following. If time and the arrangement of beds in the garden allows it, carry out the first thorough (i.e. with horn manure preparation) cultivation work about four weeks before you plan to sow, loosening the soil to a good depth. Do this on a root day. Afterwards, lightly press the surface of the prepared bed with a shovel or a board. You can also lightly smooth it with a garden roller. This seals the surface and the 'slumbering' weed seeds will start to germinate more quickly.

After around 14 days, carry out superficial cultivation with a suitable hoe, without turning over the soil, to a depth of no more than 3 to 5 cm. Even if you see hardly any weeds as yet, this work on the soil will destroy the delicate, fine seedlings of undesired plants with little effort. Now leave the bed alone for about a week until the next root day. Then apply the horn manure preparation, for instance with a hand brush, in large droplets. Once again work over the soil, though only lightly and superficially. You can now start sowing your garden crops.

Growing young plants in pots and trays

For sowing in pots or trays, of whatever kind, you need a soil that is not too rich in nutrients. This can be mixed up from two-thirds good, mature compost, around a sixth of sand and a sixth of so-called molehill earth. This is a peat-free mix in which seeds usually do well. You can also apply horn manure preparation.

The pot should naturally have holes at the bottom and stand in an under-tray or saucer so that excess water can run away – in case you water a good deal.

Sowing with cosmic impulses

The pots or trays are always freshly filled with earth just before sowing. By moving the soil around before sowing, the cosmos can act on it. The greatest possible cosmic influence occurs when the soil is moved or agitated. Merely sowing seeds is not enough for the cosmic impulses to properly work.

After sowing, the seeds are covered with fine soil with the aid of a sieve. Only as much is needed as will hide the seeds from view. Watering must of course be done carefully so that the seeds don't get silted up. Always cover the seed trays or pots with garden fleece, which will hasten germination and require less watering.

A good start for the seeds

Then a little patience is needed. Make sure you write the date of sowing, plant type and variety on the label and in your garden calendar, to document how long each sowing takes to germinate with different vegetables. Look at the trays every day so that you can remove the fleece at the right time. After the seeds have germinated, place the trays in as bright a place as possible so that the young plants don't have to seek the light and in the process form long, weak shoots. Short, strong seedlings are the goal.

Hardening off

If the weather allows, air the greenhouse so as to keep the temperature low. Seedlings on a window sill should be put outside if the weather is clement, and perhaps brought in again at night. Once the seedlings have grown beyond the cotyledon stage, and you can make out the first 'characteristic' leaves for this plant type, it is time to prick out the seedlings – that is, to place them in larger containers so that they can grow on speedily into strong young plants.

Growing young plants in garden beds

Instead of using seed trays, you can have a seed bed outside in the garden, especially for sowings later in the gardening year, from mid-May. Lettuce, cabbage and later leeks can be brought on in this way

Thorough soil cultivation work accompanied by the horn manure preparation is reinforced by a large quantity of mature compost. For sowing itself, choose as always a day that corresponds to the type of plant — a leaf day for the crops mentioned above (lettuce, leeks, cabbage).

Getting off to a good start

Sow the seeds carefully and not too close together so that the young plants have enough space to develop. The seeds also get off to a good start if carefully watered and then immediately covered with a special 'early harvest fleece' intended for seeds. Don't forget to mark the bed with a label bearing all important data that you need for compiling your experiences and making use of them in future.

Replanting

Here again, one needs to wait patiently after sowing — but keep checking what's happening under the fleece. The

Using a small weeding tool in the seed row. This removes the first weeds and thus gives the crop seedlings a good start.

moment the seeds germinate, remove the fleece in the evening so that the seedlings can habituate themselves to the harsh reality of the night-time, yet do not burn up in the sun during the day. Continue to keep an eye on your small charges. Once the young plants are big enough, they are dug up with their root balls and replanted in the bed planned for them.

The new bed will also have been thoroughly cultivated at the right constellation for the future crop. Now the plants can get growing properly.

Planting in the garden

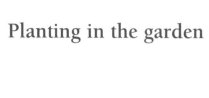

Having lovingly prepared your soil as described on page 83, you can plant out your crops.

Keeping to your crop rotation

Have you compared your current bed plans against what you did last year, to avoid planting the same type of plant in the same bed? Such a monoculture will eventually weaken the plants and make them more susceptible to disease. In

When the moon descends from the zodiac sign of Twins to Archer, it is planting time. Now plants can be sown or planted on the right day for them – blossom, fruit, leaf or root days.

90

other words, don't plant cabbage in a bed where it grew last year. There are however plants that will 'forgive' being planted in the same bed several years running, such as tomatoes and potatoes. But in the case of the cabbage family, you should leave at least four years between plantings, as otherwise you risk the dreaded plant disease club root. This is caused by a parasitic, unicellular organism that attacks the roots of most Cruciferae, including crops such as cabbage, rape, mustard, white radish, decorative plants and a large number of weeds. Club root can remain in the soil for up to 20 years, and is therefore the most feared disease in the domestic garden. In our market garden too it is the disease that impacts most severely on yield. It is caused by failing to observe a correct crop rotation plan, and can thus be avoided.

Planting depths

When planting lettuce, it is important to place them very close to the surface: more or less just on top of the soil with their root balls. This means that the wind can blow under the lettuce head,

thus helping to avoid mildew and decay.

In contrast to lettuce, leeks and cabbages are planted deep. Celeriac, on the other hand, must be planted right on the surface to help the roots develop well.

When planting crops, take care to observe the right planting depths. To be able to harvest white stalks, leeks are placed deep in the soil, whereas, to form fine corms, celeriac is planted at a very shallow depth.

Cultivating the soil

Good hoeing practice

Good hoeing accompanies plants on their way to becoming tasty vegetables or beautiful blossoms. If you hoe at the right constellation (e.g. carrots on a root day), this involves the cosmos in helping to promote good plant growth. By this means you can also improve 'planting errors' (plants sown at the wrong constellation).

Non-root weeds are, for instance, gallant soldier, orache, chickweed, speedwell and other two-cotyledon plants. They can usually be recognized by their two cotyledons and their branching roots.

These weeds are hoed out with as little earth attached to them as possible, so that they dry much faster. Shallow hoeing does not 'waken' slumbering seeds deeper in the ground.

Our crops often compete for water and nutrients with undesired weeds. Hoeing as they germinate makes weeding easier later.

Flat hoeing of weed seedlings
Hoeing in general is only done to a very shallow depth, i.e. no more than 2 cm. This cuts off all non-root weeds (those which do not propagate via their roots) so that they cannot continue growing.

Pendulum hoe with long handle. This gives an optimum standing position when hoeing and weeding, allowing both pulling and pushing movements.

Digging deeper for root weeds

Unlike the non-root weeds, root weeds such as thistle, couch grass, bindweed, buttercup and ground elder must be pulled out from a greater depth. The earth is shaken off the roots and the plants are laid in the sun to dry, then chopped up and added to the compost.

Working with the pendulum hoe

In our market garden we have achieved the best hoeing results over many years by using the 'System Glaser' pendulum hoe (see p. 123). With a long, oval ashwood handle designed by myself (see p. 122) this allows good, flat hoeing and therefore makes it easier to work for a long time without tiring. With all crops and every soil type, we use this indispensable tool almost all the year round.

The pendulum hoe allows you to work with both pulling and pushing movements. Unlike traditional hoes, the blade, lightly curved and mounted at a horizontal angle to the ground, is always in the soil when being used, and always works in two directions.

This tool comes in different blade-sizes: 85 mm for very small gaps

Draw hoe with special draw hoe position. Equipped with a long handle, this also allows an optimum upright standing position.

This position should definitely be avoided when hoeing and weeding. Bending over to work is bad for the back and will quickly tire you.

Hoeing can save the watering can

Besides getting rid of weeds, hoeing does another important job: destroying fine soil capillaries which otherwise allow the water in the soil to evaporate. Thus soil moisture is retained. This is why an old gardener's saying runs: 'Tickle the earth with a hoe and she'll laugh with a harvest.'

between plants, 125 mm for gaps up to about 20 cm, and 175 mm for gaps over 25 cm. It is important to note that the handle has to be at least 170 cm long to allow an upright hoeing position. The oval handle we ourselves have developed requires much less energy for sideways pendulum hoeing than the normal round handle.

The circular blade is sharpened on all sides and lets you work very close to a crop without damaging roots or disrupting growth. When the blade goes blunt, it is easy to change.

Another useful hand hoeing tool should be mentioned here also. This is a draw hoe with a special blade position. This special mounting position makes it possible to undertake very precise,

draw hoeing in an upright stance to a maximum depth of 2 cm. The draw hoe should likewise be equipped with a long, oval handle to allow back-friendly work.

Mulching

Mulching is one of the most important methods for aiding plant growth. It involves putting organic material that is as diverse as possible on the bare ground between plants. Chopped up materials such as grass mowings rot down quickly and are absorbed by soil organisms. Coarser material remains longer as soil cover. One can use all sorts of organic remains from the

The fin hoe allows you to loosen the soil to a good depth for root vegetables.

garden to mulch with, but do not use seed-bearing plant parts.

The modern form of mulching involves the use of black plastic or black-dyed fleece to cover the ground.

Irrespective of whether you use organic materials or 'modern' ones, the aim is the same in both cases: to suppress weeds and stimulate micro-organisms in the soil. These can only thrive in the dark, and are destroyed by sunlight. Thus mulching with organic materials at the same time supports the soil's nutrient network. It cannot however be used as a substitute for regular additions of compost.

Organic plant remains from bushes and trees, as well as grass mowings, are good for making a mulch. The materials must be chopped up well, however, before they can help improve the soil.

Plant care and attention

Some types of vegetable require further attention as they grow. Wherever possible such work should be carried out at the right constellation for the type of plant.

Pinching out tomato side-shoots

Side-shoots must be removed from tomatoes to ensure a good harvest. The side-shoots are broken off close to the stem as soon as they have reached a length of about 10 to 15 cm. This helps the fruit to develop better. In August the tomato plant is topped to prevent further flowers and fruit forming. The shoot beyond the last blossom is cut off so that the fruits already formed can ripen well: all the plant's strength goes into what is already there rather than into new growth.

Topping is done in August since the period from pollination to ripe fruit takes about 90 days.

Bell pepper: less is more

When growing bell pepper plants (capsicum) and aubergines, it is important to break off the so-called 'king's flower', or the first blossom that forms. Doing this leads to better, stronger growth at an early stage. After this first blossom is removed, the pepper plant builds up more leaf mass so that subsequent fruit will grow much better. If you don't do this, fruit form more quickly but develop less well.

Courgettes: male and female flowers

In the case of courgettes or zucchini it is possible to control the extent of the crop during blossoming, or reduce it in the holiday period. Shortly before you go away on holiday, all female flowers are removed — these are the ones that show a small fruit from the outset. Male flowers on the other hand only have a flower stalk and no fruit swelling.

Plant habitats for garden creatures

The autumn too is an important season for garden work. In the old days people would say that the garden needed to be 'made ready' for the winter. Things were cleaned up and put away, and much organic material was put on a heap in the garden or a communal compost pile.

Please don't do this! Instead leave faded flowers and dried vegetation standing in the garden over winter. Not only will they look very beautiful or strange in a hard frost, but the seeds of faded blossoms are an important food source for birds that overwinter with us, such as the robin and bullfinch. A great many insects, too, that have helped to keep this or that 'pest' in check will find winter quarters in hollow, dead stalks. These important helpers would be removed in any clean-up operation.

Leaves — a natural protection

Likewise, many think that fallen leaves are a nuisance in the garden. But wherever possible, leave them where they fall, since they offer a natural protecting cover for the ground. Leaf-fall on the lawn can simply be chopped up with the lawn mower and added to the compost heap. Take care to ensure that the heap is not filled with too many leaves at once, but in stages along with chopped prunings and some fresh organic material, for instance from the kitchen. Then active decomposition will quickly occur and after a few days the compost container will have room again for more.

The use of leaf blowers and leaf hoovers is often tempting for gardeners with a love of things technical — but they have no place in a natural garden!

The female courgette flower forms fruit.

The male courgette flower with a long stalk provides the pollen for fertilization.

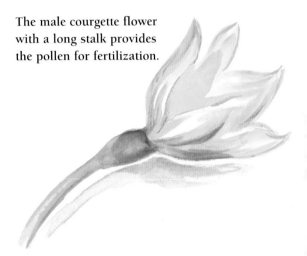

Remember that besides all the noise, many small and important garden creatures will be unsettled or killed. You wouldn't like it either if a hurricane suddenly blew you out of your garden...

Your garden is a habitat for many visible and invisible helpers.

Practical Plant Protection

Preventing insect attack

If plants are attacked, the gardener first ascertains whether this is due to animal pests or plant diseases. In both cases prevention is aided by choosing the right site for the plants, a suitable variety, the best possible soil preparation and compost application, and by using only minimal (organic!) herbicides or pesticides.

There are diverse herbal ways of protecting our crops from pest attacks, and strengthening them. For instance, if the plant is treated with a stinging nettle extract, the cabbage white butterfly will avoid laying its eggs there. Untreated plants are more susceptible.

Supporting beneficial creatures

By having blossoming plants in your garden you support important, beneficial creatures and help prevent pest attacks. When choosing flowers to grow, ensure you select ones that really do provide nutrition for the insects. Avoid 'double' flowers that have only

The basis for plant protection

Herbicides or pesticides always involve an intervention in the garden's living network. The best rule of thumb is therefore to avoid these and instead to use all measures to encourage strong plant growth from the beginning. These include: choosing the right cultivation site and time, choosing a suitable variety for your conditions, preparing the soil as well as possible and adding compost. It is also important to ensure you have a diversity of varieties, and to plant blossoming plants in as many places as possible.

very little nectar and pollen and therefore cannot offer much insect food.

Beneficial creatures are those that feed on other insects which do damage in various ways, thus keeping their population in check. The following are a few of them: ladybirds (and especially their larvae), lacewings, hoverflies, ground beetles and all the different types of parasitic wasp.

Cold extracts and liquid manures to combat pests

If the measures above have not been successful, and the beneficial creatures (given enough time) have not adequately controlled a particular pest (say aphids on your broad beans) then there are various strengthening 'teas' that will help without harming.

Cold extract of stinging nettle against aphids

Make a 24-hour cold extract of stinging nettles. Take fresh nettles (*Urtica dioica*) before they flower, chop them up small and put them in a container (of any material) with a lid. These fresh plant parts are crushed down a little and the container is filled to cover them fully. Close the container so that no creatures, such as flies looking for a place to lay their eggs, can get in. After about 24 hours, draw off this nettle water and use it to spray affected plants until the drops wet them fully, including the underside of leaves. This application is carried out on a leaf day.

The silicic acid released from the nettles is the herbal ingredient that 'annoys' the aphids and drives them from your beans without killing them. This will not rid you of this pest entirely, but that is not the prime purpose. At the same time you will strengthen the plants with this tea and enable them to gather adequate powers of resistance.

The important thing about the 24-hour nettle extract is that it must always be remade fresh if it is to protect plants. You can't store it for subsequent use. It is therefore good to have some patches of nettles growing in the garden. You will then always have an 'emergency' supply available, and will also benefit butterflies – which are very good for the ecosystem but are increasingly hard pressed today. Stinging nettles are an important food plant for the caterpillars of the small tortoiseshell, red admiral, painted lady, peacock and comma butterflies.

Stinging nettle liquid manure to strengthen plants

If you can't immediately use up all of the cold extract, you can make a liquid manure from it that will help strengthen plants and prevent susceptibility to attack.

Stir the remaining fluid vigorously, if possible every day for about two weeks. You can then use it as a liquid manure. The manure is diluted in a 1:10 ratio and poured on the ground around plants. For plants with a longer growing period such as tomatoes, the application is done every two weeks during the main growing period. In the case of quicker crops such as lettuce, just do it once. To improve the liquid manure still further, the compost preparations can be added to it.

Stinging nettle-comfrey liquid feed for strengthening plants

Another liquid manure that likewise helps strengthen plants is made of stinging nettles and comfrey. This is well known for good reason, for it contains much nitrogen, potassium, phosphorus and trace elements – therefore a fine plant strengthener.

What are herbicides and pesticides really?

All commercially available substances for combating plant diseases and pests, including organic ones, must (by law) be labelled 'herbicide' or 'pesticide'. Potash soaps (really just a kind of soft soap) for combating aphids are therefore also termed 'pesticides' although they do no harm to beneficial insects and have no side effects – unless you have a soap allergy, or drink two litres of the stuff!

Take stinging nettles and comfrey (*Symphytum officinale*) and chop them up. As with the stinging nettle feed, put these in a container with a lid.

Liquid manures need oxygen to form properly. Stir the fluid every day therefore, until the plants have brewed fully. As soon as the liquid no longer foams, after about two weeks, the liquid manure is ready for use. Add a little rock powder too, to lessen the somewhat strong smell.

Commercial products to combat aphids

If all else fails, you can use a commercial treatment which we turn to in our market garden whenever – after sufficient time and observation – the other approaches are unsuccessful. The product we use is Neudosan insecticidal soap, which is a concentrate with natural ingredients for targeting soft-bodied insects and mites on fruit, vegetables and ornamental plants both in the open and under glass. These insects include blackfly and greenfly, psyllids, whitefly, red spider mite and green spruce aphid. The treatment does no harm to ladybirds, lacewings, parasitic wasps and parasitic mites. Plants are thoroughly sprayed with a 2% solution when any attack begins.

Stinging nettle liquid manure can be made in a simple container, such as a wooden barrel. The brewing process takes about two weeks. To ensure that insects and other small creatures do not get into the container, it must definitely be sealed with a lid during this period. You will need to stir every day to add oxygen.

Slugs and snails

A plague or a help? No doubt you will be wondering how on earth these 'critters' can be called helpful, since they devour your lettuces all the time!

But, with their mucous slime, they also provide proteinous substances that micro-organisms need to break down the most diverse soil constituents. Sometimes, however, it just takes too long to establish a harmonious balance with slugs, and there are various emergency measures you can take.

Unhelpful measures against slugs

Slicing slugs in half does not strike me as a very helpful way of holding back these creatures. The carrion this leaves behind entices still more slugs, since some types of slug eat it.

Likewise it seems to me that 'slug manure' is counter-productive. This is made by pouring boiling water on slugs and then using the liquid to water vegetable beds.

Beer traps are often recommended. In my experience, though, these are not always successful compared with simply picking up the slugs. They

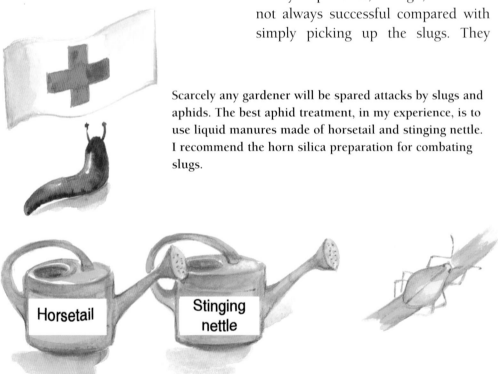

Scarcely any gardener will be spared attacks by slugs and aphids. The best aphid treatment, in my experience, is to use liquid manures made of horsetail and stinging nettle. I recommend the horn silica preparation for combating slugs.

Horsetail

Stinging nettle

only work in combination with 'slug collection' from the trap, since beer alone just draws them in larger numbers.

Barriers round the bed

Slug barriers are very useful for keeping slugs away from a bed (you can buy them commercially in various materials). However, this presupposes that you have first removed all slugs from the bed and have not overlooked a slug colony – and in fact shut them in with the barrier. There are small types of slug that often congregate just under the surface of the soil, and can easily be overlooked.

Combating slugs by hand

Collecting slugs is a much better way of proceeding, as long as you have a solution for where to take them afterwards. Thoroughly soak a board in water and place this between garden crops. The slugs will crawl under it during the day and the next morning you can collect them up. Repeat this on successive days, depending on the quantities of slugs you are dealing with. You can take the slugs to the woods, or pour boiling water over them, leave them standing in water, and afterwards bring them to the compost heap – where other creatures will enjoy the feast.

Iron-based slug pellets

If you think you need to launch a full-scale attack on slugs, use a slug killer based on ferric phosphate. This product kills slugs but, unlike ordinary slug pellets, does not intervene in the food-chain so drastically. Please don't use an ordinary slug killer based on metaldehyde or methiocarb! These substances endanger many slug-eating creatures such as hedgehogs etc., which can die as a result. Even birds of prey can be badly affected by eating poisoned slugs and the animals that eat them, for instance by going blind. They can also harm domestic pets.

Habitat for slug-eaters

A further important means for keeping slugs in check is to leave habitat and space for the most varied garden creatures. Hedgehogs, ground beetles, shrews, frogs and toads, lizards and various species of bird like eating slugs or slug eggs and young slugs. Today they are increasingly hard pressed for space and habitat, and love to be offered dead wood and piles of stones in your garden, as well as hedges, nesting materials, bug hotels, etc. You can intentionally construct such 'messy' places in the garden, then leave them undisturbed. Within a short time, hedgehog and Co. will thank you by moving in, and you will promote a balance between 'eating and being eaten'.

A cunning way to foil slugs

If you have a larger garden or even a vegetable field surrounded by meadows or rough ground, slugs will be certain to find their way in.

In this case, use a different tactic: plant the edge of the bed bordering the meadow with the double number of plants. The slugs can eat themselves full there while the other beds remain untouched. To save the doubling of costs for vegetable seed, a 'slug barrier' can also be sown – plants which slugs avoid, chiefly types of mustard. To ensure the slugs really are kept out, this barrier should be at least 50 cm wide.

Compost to combat slugs

There's a natural way to suppress slug activity, which is to regularly apply a dose of your lovingly prepared compost. This harmonizes soil processes. By increasing the garden soil's humus content, in our experience the 'slug plague' slowly but consistently diminishes.

A biodynamic slug treatment

In my view the best means of all, which comes from the repertoire of biodynamic methods, is the horn silica preparation. In this case it is sprayed directly on the soil on flower or fruit days, on the beds and between the plants. This increases light intensity in the ground and slugs don't like it at all, avoiding this area for a certain period.

If you have a bad plague of slugs, the horn silica preparation should be repeated several times on flower days.

Creating the best conditions

To be honest, even strict adherence to all this advice, along with use of biodynamic preparations, will not guarantee that plants remain absolutely pest-free. There are so many factors today that make life difficult for our crops, such as pollutants in soil, air and water, acid rain and increased ozone levels. Nevertheless, the measures described here offer the best conditions for growth and strong, resistant plants. With our care and concern they will thrive as well as they can!

Recognizing and fighting fungal diseases

Garden plants can be attacked by the most varied bacterial or viral plant diseases. Two important fungal diseases are described below, with effective methods for fighting them.

True mildew

True mildew is a fungus growing on the surface of its host plant. On the surface of a leaf a fungal network forms, with a white plaque that can be rubbed off. By drawing off nutrients, the fungus makes the leaf wither and die. True mildew overwinters as mycelium in the buds of affected plants, and starts spreading again when the plant puts forth new shoots. Wind-borne spores form additional new infection spots. The fungus affects numerous vegetables and fruit such as cabbage, salsify, lettuce, spinach, onions and grapes.

False mildew

In so-called false mildew, a greyish-blue fungus forms on the underside of leaves. Since the mid-nineteenth century it has been one of the most feared plant diseases in Europe, in both agri-culture and domestic gardening. The loss of nutrients makes leaves turn yellow and fall off. False mildew spreads primarily under moist and warm conditions in beds and green-houses. This type of fungus includes potato and tomato blight.

Horsetail tea to fight mildew

To combat attacks from both types of fungus, make a hot concoction of horsetail (equisetum). Place around 200 g horsetail (fresh or dried – available from chemists) in about 2 litres of water on the stove at a low heat, and simmer for an hour. This releases the silicic acid contained in the plant. Following 1:10 dilution with water in the early morning – preferably on a leaf day – this can be sprayed on plants. Treat the underside of leaves in particular.

This concoction of horsetail can be stored for a fairly long period (around three months) in a sealed bottle, in a cool, dry place.

For fungal diseases, however, advance preventive treatment is almost always the only really satisfactory approach.

Rich Harvest

Harvesting at the right time, and storage

There's always something to harvest in the garden, as long as your plans and actual cultivation have made sure of it. If there is suddenly a lot to harvest, such as dwarf beans whose optimum harvesting period is limited to only a few days, they should be picked in a good, ripe condition, blanched and put in the freezer. In such an instance I'm less concerned with the constellation than with picking the delicate beans at the right moment.

For harvesting purposes, cabbage plants (with the exception of broccoli) are regarded as root plants along with root vegetables, and harvested for storage on root days (preferably during non-planting time). If produce is to be eaten quickly, you can ignore the constellation.

Storing the best

In the case of vegetables which we wish to keep in storage for a longer period, such as potatoes, carrots, beetroot and other root vegetables, I always take careful note of the 'star calendar' and make sure – as long as the weather allows – that they are harvested on root days. In our market garden we have found that it is very helpful to treat such vegetables with a horn silica preparation around four weeks before harvesting, again on a root day towards evening. If you do this on three successive evenings, it can have a very positive effect on taste and storage quality. It is evident that only the best-quality produce is stored. Vegetables with small defects are eaten quickly, to prevent possible contagion in storage and greater losses. It is also important that storage vegetables are not washed or 'scrubbed clean'. Leave a little earth on them since this offers good protection against drying out.

When harvesting produce for eating fresh don't worry about the constellation. If it is for storage, however, it is best to harvest on fruit or root days (even for leaf crops such as cabbage).

110

Swiss chard is a leaf plant and for fresh consumption can be harvested on leaf or other days (preferably during non-planting time). However, for storing, it should be picked on a root or fruit day.

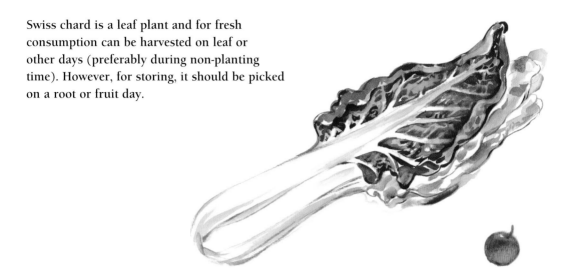

Root cellars

There are various ways of storing vegetables, and it is of course up to you to decide which method suits you best.

Root cellars are probably the most laborious means, but vegetables stay fresh for longest in them – if you care for them properly. Even in winter you can't forget about them, but must always air them a little on frost-free days, and shut them up again of course in frosty conditions. Exposure to frost means certain death for stored vegetables. Root cellars are best for root vegetables, though cabbages can be stored in them under certain conditions. They are not suitable for fruit.

Get digging

To make a root cellar, dig a pit about 50 to 70 cm deep. The soil you excavate is placed around the pit as raised wall. You can ensure mice don't spoil the harvest by lining the pit with walnut-tree foliage, into which you place the vegetables. As mouse precaution, take care that soil and walls of the cellar have a sufficiently thick layer of foliage, and that the stored crop is kept about 10 cm away from the wall of the pit. If you can't get hold of walnut-tree foliage you can seal the cellar against mice with narrow-meshed rabbit-hutch wire, with which you line the bottom and walls of the pit. Finally, cover the cellar with sufficient foliage again or with the wire mesh, and place several boards over the top. Then insulate it with bubble wrap or straw, so that neither frost nor heat can enter too strongly. If the weather is warm for several consecutive days, the cellar must be regularly aired.

Instead of foliage or wire mesh, you could also bury an old washing machine drum in the garden, preferably a top-loader with press-down lid. If you only have a front loader, use a flagstone as lid.

Safe in your own cellar

If you have a cellar in your house with an earth floor, you can easily store your vegetables there. But again, do not forget to air the cellar regularly on frost-free days so that the storage temperature stays as low as possible, preferably below 5°C. If you store vegetables in the cellar, make sure you store apples in a different room. The ripening gas ethylene, which apples emanate, considerably impairs the storability of the other vegetables. If you only have one storage room, place the apples in a bucket with a lid to protect the vegetables.

Converting the use of cold frames

You can also store vegetables in a cold frame. Cabbages, leeks, endive and sugarloaf can be 'planted back' as whole plant with the root trunk. In other words, harvest the cabbages with a small root ball and replant them in the cold frame, packed close. It is fine for the plants to touch each other. Here again you must make sure that the frame is aired on sunny days. To protect it from frost, you could cover it with bubble wrap.

For gardeners with a greenhouse

If you have a greenhouse, you can equally well store leeks, sugarloaf chicory, endives and cabbage there. Here again, you must beware of frost and, depending on the type of greenhouse, possibly seal it with bubble wrap. On sunny spring days it will be important to ventilate the greenhouse, for the first rays of sun will warm it up quickly, which would be bad for the vegetables.

A cold frame for storing crops. Walnut-tree foliage and a container with a lid offer good protection against mice.

Simply in a sand box

If you don't want to use any of the above methods, you can put root vegetables in layers in a box in your cellar, with sand between every layer. On top of the last layer of sand place foil to ensure the vegetables don't dry out. You will need to check the box regularly, monitoring the moisture and when necessary remoistening the sand.

Hanging vegetables on a line

Many vegetables can simply be stored in a 'normal' cellar as long as it is kept

Freshly harvested carrots (on a root day). They are not washed before storage in a root cellar, in a house cellar or in a sand box. The soil helps protect them from drying out.

The best sauerkraut

Lactic acid fermentation is a special way of storing vegetables. Sauerkraut is the best known form. In autumn take good heads of white cabbage. Sauerkraut specialists cultivate particular varieties such as 'Holsteiner Platter' or 'Donator'. The best sauerkraut is made by harvesting and bottling on a flower day. This is how we've been doing it for many years – with great success.

cool enough. For instance cabbage heads, tied together in twos, can be hung up by their stalks on a kind of 'washing line'. You can do the same with sugarloaf chicory. However these should still have a little root to tie them on with. In the case of sugarloaf chicory make sure that the remains of the root have no soil clinging to them, as this will impair storability.

Saving Your Own Seed

Seed diversity: the more the merrier

It's good to grow whatever you like to eat. Our gardens can become an inexhaustible source of daily food and health. As Hippocrates said (460–377 BC), 'Your food should be your medicine.'

Alongside all the well-known crops such as carrots, beetroot, lettuce and spinach, I'd like to give a few examples here of less common but tasty plants that are good to grow. One advantage of growing your own is the opportunity to have vegetables you might look for in vain in ordinary shops, with their relatively limited range – although sometimes you can find unusual foods in market gardens that have direct sales outlets. Choose diversity – the more the merrier – and just try out whatever takes your fancy.

Seed catalogues or plant descriptions in gardening magazines can offer stimulus here. Use labels, too, so that you can compile records of your experiences with different plants. You can also collect the seeds and so have your own seed stock for the following year's crops. Useful addresses of organic seed suppliers can be found on page 123.

Unusual vegetables – give them a go

Oriental salad leaves and greens
These come in the most diverse forms and colours – such as green and red mustard leaf. These two leaf crops are tasty and very robust. 'Mizuna' on the other hand has green, finely cut leaves with a mild taste. Oriental greens are planted between September and mid-May, either by sowing directly in beds or in seed trays for growing on. As oriental greens are brassicas they are often peppered with holes by flea beetles. And long, hot summer days quickly start them shooting – in other words forming seeds so that they're no longer much good for harvesting.

Catalogna lettuce
A cut-and-come-again lettuce with long, dandelion-shaped leaves, and is slow to bolt. There are both green and red leaved varieties. Sow spring to autumn.

Winter purslane (*Claytonia perfoliata*)
Also known as miner's lettuce, this is an ideal salad plant for growing during shorter, cooler days. It can be cultivated in the open, in a cold frame or in a cool greenhouse. Purslane is a good source of vitamins in the winter months.

American land cress (*Barbarea praecox*)
This is a frost-resistant, tasty salad plant that is grown in the open during winter.

Purple-blue podded peas, such as 'Ezetha's Krombek Blauwschok' or 'Blouwschokker' grow up to 2 metres high and require a support. The peas are very tasty and good for eating fresh.

Chicory varieties (*Cichorium*)

Varieties such as 'Red Verona' or 'Green Grumulo' are suitable for early harvesting in spring. They are sown from June to July, and overwinter on the bed. Frost protection using fleece is a good idea, and also brings forward the spring harvest considerably.

Red orache (*Atriplex*)

This is a lovely plant whose leaves are also decorative as well as being a good spinach substitute. Good King Henry (*Chenopodium bonus-henricus*) is a relative, now hardly known, but in past times grown widely and used like spinach.

Garden sorrel (*Rumex acetosa*)

No herb garden should be without this plant, which can be added as extra spice to salads.

Scorzonera (*Scorzerona hispanica*)

This is also known as black salsify or oyster plant. It has a distinctive taste and requires good, deep soil preparation to produce strong roots that can be peeled. The whole root is taken from the earth.

Sima pea (*Pisum sativum* spec.)

This is an overwintering pea, compact and very well suited for garden growing. It is ready to harvest in the spring from an October sowing.

Jersey kale (*Brassica oleracea* var. *palmifolia*)

Also called 'palm kale' or 'walking-stick kale' this is very similar to green cabbage but does not overwinter. The variety 'Nero di Toscana' can be planted from the beginning of May and is a welcome addition to your vegetable range. In general, Jersey kale is milder in taste than green cabbage. Finely chopped, the leaves can be used in salads or for cooking. It can also easily be blanched then frozen for later use, and is excellent for Tuscan winter stews. One well-known recipe that uses Jersey kale is the Portuguese cabbage soup 'caldo verde'. It can be grown, too, as a decorative tub plant.

In former times, the long, woody shoots of this plant were used for making walking sticks.

A touch of holiday atmosphere

The list of interesting vegetables could go on and on. On holiday abroad one can get good ideas. I make a point of going to places where seeds and plants are sold and buying varieties I find there, adding new, interesting vegetables and lettuces to our range, to our customers' delight. Why not try out something new and see what happens?

Organic-quality seeds

Whenever possible buy seeds from organic or biodynamic cultivation. Then you will know that the seeds come from stock that was grown in harmonious interplay with the soil's nutrient network, and will be predisposed to seek out nutrients that enable them to grow into healthy food.

Avoid using hybrid seeds. Stock grown from them often cannot reproduce in subsequent years, or will alter greatly from the original plants. For instance, very divergent cabbage heads will grow.

To harvest seeds from your own tomatoes, pollination must take place. Bumble-bees (not honey bees) or the wind undertake this work. If you wish to lend a helping hand, you can shake the plants lightly.

Avoiding hybrids

You can always grow your own seed on blossoming vegetable plants as long as you haven't sown hybrid varieties. Hybrids often cannot reproduce further, or are unable to pass on the desired qualities to the next plant generation. You can see on the seed-pack label whether the variety is a hybrid. In this case, alongside the name of the variety, you will usually find the addition 'F1' (F stands for *filia* or 'daughter' in Latin).

When buying young vegetable plants, don't forget to ask what the variety is. You can then put all relevant details on your labels.

Seed-growing for beginners

If you wish to harvest your own seeds, it is best to start with dill, borage or cress – for these three lend themselves very well to seed-reproduction trials. They're almost always successful, and the flowers of these plants also give especial pleasure to many useful insects.

I would not try harvesting seeds from pumpkins and squashes for these cross with all other relatives, such as decorative squashes and courgettes, producing the most bizarre results. It's a shame if cross-pollination by a decorative squash leads to a bitter-tasting pumpkin the following year.

119

Assigning garden plants to different groups

Below is a short survey of garden plants as assigned to their different groups. The group is usually determined by the part of the plant that will be harvested or used, and once you get the hang of this you can easily decide which group each plant belongs to.

Root plants

Beetroot, carrots, celeriac, chicory root, garlic, horseradish, Jerusalem artichoke, onions, parsnip, potatoes, radish, scorzonera, sweet potatoes

Leaf plants

Basil, Batavia lettuce, borage, Brussels sprouts, butterhead lettuce, cauliflower, celery, chard, chervil, chicory, Chinese cabbage, chives, cress, endives, fennel root, garden orache, green cabbage, iceberg lettuce, kohlrabi, lamb's lettuce, lawn grass, leeks, lemon balm, New Zealand spinach, pak choi, parsley, 'pick-and come-again' (looseleaf) lettuce, radicchio, rhubarb, rocket, Romaine lettuce, savory, Savoy cabbage, spinach, sugarloaf chicory, white cabbage

Flower plants

Artichokes, broccoli, chamomile, cutting flowers, flower bulbs, flowering shrubs, lavender, roses, summer bedding flowers

Fruit plants

Apples, apricots, aubergines, blackberries, blackcurrant, blueberries, cherries (sweet or sour), courgettes, cranberries, cucumber, dwarf beans, figs, gooseberries, grapes, greengages, hazelnuts, Japanese wineberry, kiwis, loganberries, maize/sweetcorn, melons, mirabelles, nectarines, peaches, pears, peas, peppers (capsicum), plums, quince, raspberries, redcurrant, runner bean, soya, squashes/pumpkins, strawberries, tomatoes, walnuts

Cultivation work at planting time and non-planting time

During planting time	During non-planting time
Fruit plants	*Fruit plants*
Sowing, planting out and replanting of fruiting vegetables and fruit bushes	Cultivation work above ground, e.g. pest and disease treatments (not hoeing or pruning)
Giving compost and manure	Taking cuttings from fruit trees/bushes
Hoeing, rooting cuttings	Taking grafting scions from fruit trees
Flower plants	*Flower plants*
Sowing, planting out and replanting of flowers, blossoming herbs, blossom vegetables (e.g. broccoli) and decorative trees/bushes planted for their flowers	Cultivation work above ground, e.g. pest and disease treatments (not hoeing or pruning)
Cutting flowering shrubs and hedges	Taking cuttings
Giving compost and manure	Cutting grafting scions from flowering bushes/ shrubs
Hoeing, rooting cuttings	Harvesting, also for drying
Planting flower bulbs	Harvesting of leaf vegetables and leaf herbs*
Mowing grass if it is to grow slowly	
Leaf plants	*Leaf plants*
Sowing, planting out and replanting of leaf vegetables, leaf herbs, decorative foliage plants, conifers, hedges	Cultivation work above ground, e.g. pest and disease treatments (not hoeing or pruning)
Cutting back broad-leaf trees, conifers and hedges	Taking cuttings
Giving compost and manure	Cutting grafting scions from bushes/shrubs grown for decorative foliage
Hoeing, rooting cuttings	Harvesting
Fertilizing lawns and grass meadows	Mowing grass if it is to grow quickly and thickly
Root plants	*Root plants*
Sowing, planting out and replanting of root vegetables and root herbs	Cultivation work above ground, e.g. pest and disease treatments (not hoeing)
Giving compost and manure	Harvesting
Hoeing, harvesting of root vegetables	

* Maria Thun's research has found this is best done on flower days

121

Useful addresses

For information on specialist gardening implements etc.:
Peter Berg
Gärtnerei Berg
Niederfeld 1
79589 Binzen
Germany
Phone: (49) (0)7621 96 83 10
email: *info@bergbinzen.de*
www.bergbinzen.de

The worldwide biodynamic movement is represented by:
The Agriculture Section
Goetheanum
CH-4143 Dornach
Switzerland
Phone: 41 (0)61 706 4212
email:
sektion.landwirtschaft@goetheanum.ch
www.sektion-landwirtschaft.org

For contacts worldwide:
Demeter International
email: *info@demeter.de*
www.demeter.net

UK:
Biodynamic Agricultural Association
Painswick Inn Project
Gloucester Street
Stroud, Glos. GL5 1QG
Phone: (44) (0)1453 759501
email: *office@biodynamic.org.uk*
www.biodynamic.org.uk

USA:
Biodynamic Farming and Gardening Association
25844 Butler Road
Junction City
OR 97448
Phone: (1) (262) 649 9212
email: *info@biodynamics.com*
www.biodynamics.com

Contact the above for advice on biodynamic trainings and literature.

For rock dust:
Seer Centre Trust
Ceanghline
Straloch Farm
Enochdhu
Blairgowrie
PH10 7PJ
Scotland
Phone/Fax: (44) (0)1250 881789
www.seercentre.org.uk

The Organic Gardening Catalogue
Riverdene Business Park
Molesey Road
Hersham
Surrey
KT12 4RG
Phone: (44) (0)1932 253666
www.organiccatalogue.com

USEFUL ADDRESSES

For pendulum hoe:
Glaser Engineering GmbH
Im Lerchengarten 12
CH-4153 Reinach
Switzerland
email: *info@glaser-swissmade.com*
Phone: (41) (0)61 713 98 84

Biodynamic seed suppliers in the UK and Europe:
Stormy Hall Demeter Seeds
Botton Village
Danby, Whitby
North Yorkshire
YO21 2NJ
Phone: (44) (0)1287-661368
email: *stormy.hall.botton@camphill.org.uk*

Poyntzfield Herb Nursery
Black Isle, by Dingwall
Ross-shire
IV7 8LX
Scotland
Phone: (44) (0)1381 610352
email: *info@poyntzfieldherbs.co.uk*
www.poyntzfieldherbs.co.uk

Sativa Seeds
Sativa Rheinau AG
Klosterplatz
CH-8462 Rheinau
Switzerland
Phone: (41) (0) 52 304 91 60
email: *sativa@sativa-rheinau.ch*
www.sativa-rheinau.ch

Bingenheimer Saatgut AG
Kronstrasse
2461209 Echzell-Bingenheim
Germany
Phone: (49) (0) 6035/1899-0
email: *info@oekoseeds.de*
www.oekoseeds.de

Reinsaat
A-3572 St.
Leonhard am Hornerwald
Austria
Phone: (43) (0) 2987 2347
email: *reinsaat@reinsaat.co.at*
www.reinsaat.co.at

For information and resources on organic gardening and agriculture:
The Soil Association
Bristol House
40–56 Victoria Street
Bristol BS1 6BY
www.soilassociation.org

Garden Organic
Ryton
Coventry CV8 3LG
UK
www.gardenorganic.org.uk

Soil analysis services in the UK:
The Royal Horticulture Society
80 Vincent Square
London
SW1P 2PE
Phone: (44) (0)845 260 5000
Email: *gardeningadvice@rhs.org.uk*
www.rhs.org.uk

Reference literature:
Maria Thun: *The Biodynamic Sowing and Planting Calendar* (Floris Books)
Maria Thun: *The Biodynamic Year: increasing yield, quality and flavour, 100 helpful tips for the gardener or smallholder* (Temple Lodge Publishing)

Index

INDEX

Acknowledgments

Writing a book alongside the daily work of garden manager was not easy. I was only able to do this because the people around me showed great understanding for the endeavour.

My wife Christina, especially, had to pick up the pieces frequently, even more so than usual. I owe thanks to my sons Andreas and Stefan, since they took on some of the daily running of the garden while I was writing. When a gardener writes a book it is important to have someone who can scatter a handful of commas at the right places, and smooth out the grammar. Heike thoroughly supervised the manuscript's development, and made it more readable. By asking questions from a lay-person's perspective, she also greatly helped to make it accessible for those new to gardening.

For many years, Floor and Gerhard Eisenkolb have been at my side to help and advise, and are important discussion partners when it comes to both gardening and spiritual matters.

Dr Hans Balmer from Basel – a 'high priest of compost' – responded fulsomely to me regarding questions of proper, active, aerobic composting, and its more esoteric dimensions.

I would like to thank Frau Helga Reimold, as representative of all the members of the garden group at the Demeter training and show gardens, for our lively exchanges and continual readiness to address important gardening issues.

Translated by Matthew Barton

Temple Lodge Publishing
Hillside House, The Square
Forest Row, RH18 5ES

www.templelodge.com

Published by Temple Lodge 2012

Originally published in German under the title *Der Mondgärtner: Biodynamisch gärtnern mit Peter Berg* by Franckh-Kosmos Verlags-GmbH & Co., Stuttgart

© Franckh-Kosmos Verlags-GmbH & Co. 2011

This translation © Temple Lodge Publishing 2012

With 27 colour photos of Gärtnerei Berg, Binzen, and 110 illustrations by Marie-Laure Viriot, Sipplingen

The author asserts his moral right to be identified as the author of this work

A catalogue record for this book is available from the British Library

ISBN 978 1 906999 37 7

Cover by Andrew Morgan Design
Typeset by DP Photosetting, Neath, West Glamorgan
Printed and bound by 1010 Printing International Ltd., China